HYPERTHERMIA AND ONCOLOGY
Volume 1

HYPERTHERMIA AND ONCOLOGY
Volume 1

Thermal effects on cells and tissues

Edited by

M. Urano
Massachusetts General Hospital, Boston, USA

and

E. Douple
Dartmouth–Hitchcock Medical Center, Hanover, New Hampshire, USA

CRC Press
Taylor & Francis Group
Boca Raton London New York

CRC Press is an imprint of the
Taylor & Francis Group, an **informa** business

First published 1988 by VSP BV

Published 2021 by CRC Press

Taylor & Francis Group
6000 Broken Sound Parkway NW, Suite 300
Boca Raton, FL 33487-2742

ISBN 13: 978-90-6764-087-9 (hbk)

Visit the Taylor & Francis Web site at
http://www.taylorandfrancis.com

and the CRC Press Web site at
http://www.crcpress.com

CIP-DATA KONINKLIJKE BIBLIOTHEEK, DEN HAAG

Hyperthermia and oncology/eds.: M. Urano and E. Douple — Utrecht : VSP
Vol. 1: Thermal effects on cells and tissues. — Ill.
With index.
ISBN 90–6764–087–5 bound
SISO 605.91 UDC 612.014.4:616–006.6
Subject heading: hyperthermic oncology.

Contents

Hyperthermia and Oncology, Vol. 1, pp. 1–12 (1988)
Urano and Double (Eds)
© 1988 VSP.

Chapter 1

Prospects for hyperthermia in cancer therapy

J. ROBERT STEWART
Division of Radiation Oncology, University of Utah Health Sciences Center, Salt Lake City, UT 84132, USA

A. INTRODUCTION

It is appropriate that this series of volumes on hyperthermia should begin with one dedicated to an in-depth exploration of the biology of hyperthermia. This is so because the biological aspects of hyperthermia provide a firm and exciting rationale for its use in the multimodality clinical care of patients with cancer. On the other hand, it is appropriate also that the introductory chapter in the series would be authored by a clinician. The scientific history of hyperthermia, like most aspects of medicine, can be traced to initial observations by astute and creative clinicians. Apparent spontaneous remissions of cancer were noted following prolonged intense fevers due to a variety of infections. This led Coley (1893, 1911) to devise a hyperthermia treatment method by deliberate production of fever by administering a preparation containing bacterial pyrogenic toxins. Over a period of a few decades, he described instances of often dramatic response to these toxins. Local methods of heating were also applied, often with concomitant ionizing radiation, again with some excellent responses (Doyen 1910; Crile 1962). In spite of these promising observations the modality was difficult to administer with control, responses were unpredictable, and the explanation or biological rationale for the observed effect was unclear. As more predictable treatments were developed and refined, particularly surgery, ionizing radiation, and later chemotherapy, there was little further development of hyperthermia and the approach was largely abandoned. A series of experiments in transplanted mouse tumors were described in 1963 by George Crile, Jr. (1963) in an article which in retrospect is truly remarkable. In this paper he made a number of observations which were prophetic of many biological phenomena that were quantitatively characterized much later. These phenomena included time–temperature dependent cytotoxicity of hyperthermia; delayed *in situ* increased cytotoxicity and increased sensitivity of large versus small newly transplanted tumors (now both thought to be due to vascular events);

thermotolerance of both normal tissue and tumor; radiation sensitization by heat; and the temporal decay of radiation sensitization. Subsequently, quantitative experimental approaches developed by physiologists and radiation biologists were adapted to the study of the biological effects of hyperthermia. This resulted in a rapid increase in our understanding of thermal biology which has been documented by a virtual flood of publications. With improved understanding of the biology of hyperthermia leading to a more clearly defined rationale for its clinical use, a number of technical approaches were devised for clinical application of hyperthermia and for thermometry. It is the purpose of this volume to document thoroughly and discuss the current state of our knowledge of the biological, physiological, and metabolic aspects of hyperthermia. My purpose in this introductory chapter is to attempt to relate a clinical perspective to the biological and technical aspects of hyperthermia.

B. BIOLOGICAL ASPECTS

Since the intent of this volume is to provide an extensive description of the biology of hyperthermia, my comments in this section are meant to be brief and to highlight certain aspects which appear to be relevant clinically. Firstly, at temperatures above 42 °C or 43 °C there is a time–temperature dependent cytotoxicity that can be easily measured by clonogenic assays (Dewey et al. 1977), and for each increase of 1 °C in temperature the time of exposure can be reduced by one-half to create equivalent cytotoxicity (Henle and Dethlefsen, 1980; Sapareto and Dewey, 1984). The general shape of hyperthermia survival curves is similar to that seen with radiation survival curves, having a shoulder and a log-linear exponential portion. It is possible to document several logs of cell kill by hyperthermia at progressively increasing time at temperatures above 43 °C. Of potential importance clinically is the modification of cytotoxicity by nutrient deficient, acidic conditions. Hypoxia itself has little effect on the cytotoxic response, however, lowering the pH and reducing certain nutrient levels has a marked enhancing effect on hyperthermic cytotoxicity (Gerweck 1977; Gerweck et al. 1984). Since one might expect that hypoxic and therefore radiation resistant portions of tumors would also be nutrient poor and at low pH, these effects might be important in the clinical application of hyperthermia as an agent complementary to ionizing radiation. Hyperthermia, like many other cytotoxic agents, has an age response through the cell cycle. Interestingly, the most heat sensitive part of the cell cycle is the late S phase (Westra and Dewey 1971) which generally is the most radioresistant, again suggesting that hyperthermia might be both additive and complementary to radiation therapy.

Hyperthermia has been shown also to be a radiation sensitizer with the maximum effect expressed when the two agents are administered simultaneously. The interaction decays with time and becomes unmeasurable after separation of a few hours between exposures in vitro (Dewey et al. 1977). The most optimal sequencing of the two treatments to obtain maximal therapeutic gain remains to be more clearly worked out. However, what is clear is that strategies which would depend on sensitization of radiation response would require exposures as nearly simultaneous as possible and a high degree of localization of either or both of the modalities to the tumor volume in order to decrease effects in normal tissues. On the other hand, strategies to minimize interaction in order to avoid enhanced normal tissue consequences should separate the treatments by several hours and would rely on additive and complementary cytotoxicity rather than radiation sensitization for the therapeutic gain. At dose rates commonly used for external beam radiation therapy there is little sensitization at temperatures below 42 or 43 °C. On the other hand, at low dose rates, such as might be used in interstitial or intracavitary irradiation, temperatures below 42 °C do sensitize to the effects of ionizing radiation (Ben Hur *et al.* 1974). Such temperatures are more easily achieved clinically and better tolerated by the patient than 43 °C and above, and sensitization at lower temperatures is of potential importance to combined interstitial irradiation and hyperthermia.

The various phenomena described above were all initially studied *in vitro*; however, essentially every important biological feature has also been observed in either or both tumors and normal tissues *in vivo*. Normal tissue complications have not been a significant problem clinically, but the biological studies on normal tissue effects, which are the subject of a subsequent chapter in this volume, do raise the possibility that major complications could arise in situations where we administer *effective levels* of hyperthermia to *critical normal tissues* and particularly those that are dose limiting in radiation therapy. In addition, hyperthermia has complex effects on circulation which have significant potential impact on its clinical use. Most normal tissues when faced with a heat challenge react by increasing blood flow which serves to dissipate the heat; however, these normal physiological responses are often deficient or totally lacking in the neovasculature of tumors. This has potentially great importance in terms of differential heating of normal tissues and tumors growing within these normal tissues. This would, of course, apply only to the bulk of the tumor mass; preferential heating of tumor cells infiltrating microscopically into normal tissue at the tumor periphery appears to be a physical impossibility. At high hyperthermic time–temperature combinations microvascular damage occurs which profoundly influences the therapeutic response. The microvascular effects appear to occur at lower heat doses in at least some tumors than most normal tissues, and in some experimental tumors and perhaps some clinical situations vascular collapse appears to be an important determinant of tumor regression and control. In one tumor studied, tumor control appears to be due to immediate cell kill from hyperthermia rather than delayed cell killing from microvascular damage (Wallen *et al.* 1986). Vascular damage by hyperthermia could increase the number of hypoxic cells in a tumor and therefore decrease the effect of subsequent irradiation (Song *et al.* 1982; Urano and Kahn 1983). Decreased circulatory capacity could also adversely affect access of drugs to the tumor in combined heat and chemotherapy protocols.

   The interaction of chemotherapy and hyperthermia is highly complex and incompletely studied. Studies to date have been almost entirely done *in vitro*. Pharmacokinetics, drug metabolism, access of drugs to tumors, and other *in vivo* events may be affected by hyperthermia in crucial and very complicated ways. At this point, one can say that certain of the commonly used chemotherapeutic agents, such as some alkylating agents, Cisplatin, and Bleomycin, do show enhanced effects with hyperthermia. Within this group there are some, such as Bleomycin, which show significant interaction only at temperatures above a certain threshold (Hahn 1979). Others show temperature-dependent changes at lower temperatures such as 39 to 42 °C. Examples of this category are BCNU and Cisplatin. Others such as Adriamycin have an extremely complex interaction, and some, such as Hydroxyurea, Methotrexate and Vincristine, appear to have no interaction at all. In view of the interactions at relatively modest temperatures with some chemotherapeutic agents, an interesting approach might be combined chemotherapy and whole-body hyperthermia. At this point very little in the way of good quantitative biological studies or clinical results have been presented in support of such treatment. In addition to the classical cytotoxic agents, other types of drugs have shown some interesting interactions with hyperthermia. For instance, the hypoxic cell sensitizer, Misonidazole, shows a temperature-dependent cytotoxicity in hypoxic cells which has been proposed as a potentially useful clinical strategy (Stewart *et al*. 1983). In addition, alcohol, (Li *et al*. 1980; Henle 1981) and local anesthetics (Yau 1979) increase the response to hyperthermia.

   From the above brief descriptions and the greatly augmented material contained in later chapters of this volume, one can define a rationale for the use of hyperthermia in the multimodality treatment of cancer. The rationale relies on (1) additive effects of cytotoxicity; (2) complementary effects with radiation due to differences in S-phase sensitivity and the increased effects of heat in metabolic circumstances of acidic, nutrient deficient, and hypoxic milieu that might be expected to render tumors radioresistant; and (3) synergistic effects due to modification of radiation response and the response of certain chemotherapeutic agents.

C.   TECHNICAL ASPECTS OF HYPERTHERMIA

A subsequent volume will deal with the engineering and physics of hyperthermia, its measurements and devices for clinical use; however, in order to understand the current status of clinical trials, some technical description is needed.

1.   *Methods to produce hyperthermia*
The technical approaches and problems encountered vary immensely depending upon the tumor site to be heated. For instance, the problems encountered in heating superficial tumors in the skin or subcutaneous tissues

are much different and much more simple than those encountered in hyperthermia of deep-seated abdominal or pelvic neoplasms. The methods currently in use rely mainly on ultrasonic or electromagnetic devices. A variety of ultrasound devices operating between 0.3 MHz and 6 MHz have been applied in clinical situations. There are three general approaches to electromagnetic means of producing hyperthermia: external wave applicators operating between 50 MHz and 2450 MHz, resistive capacitance heating in the low MHz or kHz range, and magnetic induction. Electromagnetic methods can also be applied to interstitial or intracavitary hyperthermia either by creating areas of localized current flow between sets of electrodes in tissue (Aristozabal and Oleson 1984) or by the placement of radiative microwave antennae (Emami *et al.* 1984; Coughlin *et al.* 1985). Heating in some special cases is achieved by other means such as limb perfusion, superficial water bath heating, or transureteral irrigation of the surface of the urinary bladder. Superficial tumors have been effectively heated using either ultrasonic or electromagnetic means and good results are reported with either approach. There is difficulty with ultrasound at bone–soft tissue interfaces where increased heating can cause pain from the periosteum and ultrasound will not traverse air-containing organs.

Treating deep tumor sites is considerably more difficult. Ultrasound approaches to deep heating have involved multiple focused transducer arrays (Marmor 1983) or a moving transducer (Lele and Parker 1982). Electromagnetic approaches lead to regional rather than local hyperthermia. At the lower frequencies needed to gain penetration by electromagnetic waves, the wavelength in tissue is long and applicators must be large in relation to the body resulting in regional heating patterns. Regional heating is produced also by magnetic induction or capacitance approaches. Magnetic induction by superficial coils results in very superficial heating, while circumferential coils, such as the Magnetrode, lead to inhomogeneous energy deposition maximal at the surface and zero at the central axis of the heated volume (Oleson 1982; Paliwal *et al.* 1982; Strohbehn 1982). Because of tissue conductive properties, capacitance heating results in maximal heating in fat in perpendicular layers such as in the subcutaneous fat layer. This can sometimes be circumvented by superficial cooling, but only when the fat layer is relatively thin (Kato *et al.* 1985). Multiple electromagnetic wave applicators can be placed in an array around the patient and operated in phase to achieve the reinforcement of heating at depth. This is the principle underlying the annular phased array (Turner 1984). In spite of rapid progress in development of a number of commercially available devices for hyperthermia, a major clinical impediment to progress remains the technical difficulty of obtaining satisfactory heating in deep tumor sites.

Sites that can be approached either through exposed cavities or at surgery can be implanted with hollow electrodes or catheters for afterloading interstitial or intracavitary treatment with both hyperthermia and radiation. These approaches have the advantages of potentially excellent localization and capitalizing on the biological interaction of low dose rate irradiation with hyperthermia. However, it must be remembered that the method requires stringent conformation to geometric principles of placement. The geometric principles for heating and for irradiation are similar but not always identical, which along with the need for extra placement of carriers for thermometry can lead to highly invasive procedures for combined treatment. It has not been clearly shown that a tumor that can be implanted with ideal geometry for radiation is any further helped by hyperthermia. Flaws in the geometric placement that result in underdosing will affect both modalities.

2. *Thermometry*

The second technical area of immense importance in clinical hyperthermia is thermometry. In order to interpret results and assess responses critically, it is essential to measure temperatures achieved in the tumor and selected normal tissue sites. Those of us who are accustomed to working with ionizing radiation with the elegant dosimetry models in use cannot help but be dismayed by the awkward and invasive approaches needed even to begin to understand 'dose' in hyperthermia. No physical unit has been agreed upon to describe dose and there are no precise models, so all treatments must be monitored by thermometry. Ideally, thermometry would be accomplished noninvasively, but since such a technique does not exist, the approaches are of necessity invasive. All available hyperthermia techniques result in considerable inhomogeneity of heating; therefore, one must monitor multiple points at multiple times over the course of the treatment in order to obtain even rudimentary information concerning the effectiveness of the heating. A method of obtaining multiple point thermometry from a single catheter, thermal mapping, has been described and a semi-automated computer-controlled device is available to accomplish the maps (Gibbs 1983). Temperature-sensing devices with multiple points are also under development and some are available clinically. These approaches lead to increasing invasiveness for thermometry and reams of data to describe the treatment. A means for reducing the sheer bulk of the thermometry data considerably and stating it in the form of 'thermal dose' has been described (Sapozink *et al*. 1985b). For ultrasound hyperthermia standard thermocouples are satisfactory but care must be used in selecting sheathing material because some plastic coatings have a high absorption of ultrasound relative to tissue which results in spuriously high readings. With electromagnetic techniques, thermometry utilizing instruments with metallic components may result in artifacts such as local tissue heating from heating of the thermometer tip, or induced interference with the metallic leads. Techniques have been devised for accurate thermometry with thermocouples or thermistors for superficial electromagnetic heating, but with deep regional methods some type of non-interactive thermometer is essential. Miniature thermistors with high-resistance carbon leads or optical systems are commercially available for this purpose. The importance of accurate and abundant thermometry cannot be overemphasized.

D. SURVEY OF CLINICAL RESULTS

1. *Superficial tumor sites*

Most reports in the literature concern results of application of hyperthermia to superficial tumor sites. Tumors in these sites, usually chest wall recurrences from breast cancer, superficial melanomas or sarcomas, or recurrent squamous cell carcinoma of the head and neck, often can be relatively easily heated by the various electromagnetic or ultrasound approaches mentioned above. Much of the reported work has suffered from incomplete description of tumor extent, incomplete description of thermometry and uncertainty as to patient selection factors. Detailed review of many of these publications was recently undertaken by Meyer (1984). Fewer studies have attempted to compare hyperthermia alone with X-ray alone, or with hyperthermia plus X-ray, and fewer yet with hyperthermia plus chemotherapy. In studies that had X-ray and X-ray plus hyperthermia arms recently reviewed by Gibbs and Stewart (1985), it was apparent that for equal doses of radiation the complete response rate in the hyperthermia arm was about double that seen for X-ray alone. A randomized study is being undertaken by the Radiation Therapy Oncology Group (RTOG) which should be analyzed and published soon. It is important to stress that not only complete response rate, but also duration of response and influence of response on survival are all important parameters in evaluating these clinical series. It appears that complete responses obtained with hyperthermia plus radiation are more durable than those obtained with radiation alone and that response rates correlate with survival (Scott *et al.* 1984). From the diverse clinical reports available on superficial heating, several conclusions can be drawn.

1. Hyperthermia does have efficacy as an anti-cancer agent in man.
2. When applied with X-ray therapy there is an increase in response rate documented in a number of series.
3. With heat plus X-ray, the complete responses are durable.
4. Hyperthermia when used alone has a low complete response rate and the responses are transient in duration. This probably results from incomplete heating of the entire tumor leaving relatively unheated nests of tumor cells which grow back quickly.
5. Although patients have been bothered by a number of transient side effects such as pain, complications have been few and predictable by heating patterns of the devices being used. In most reports, the only complications have been burns which are usually relatively minor.

Many superficial sites are treatable by interstitial means. In pilot studies so far published, it appears that both the concurrent radiation dose and the effectiveness of heating are predictors of complete response (Aristozabal and Oleson 1984). Equipment has been considerably improved for interstitial hyperthermia and interstitial thermoradiation therapy, leading to the expectation that wider use and more evaluable publications will follow soon. RTOG has recently opened a randomized evaluation of recurrent cancers treated by either interstitial radiation alone or the same radiation plus interstitial hyperthermia.

2. Hyperthermia of deep-seated truncal lesions

Achieving hyperthermia in deeply located tumor sites is considerably more complicated than for superficial sites. Not only is it more difficult to achieve the desired deposition of energy, its measurement by thermometry is also more complicated. The most frequently used device to date has been the Magnetrode. Because of the physical characteristics of the device, one would predict that the heating pattern would be mostly confined to the superficial body annulus, and that tumors central in the pelvis or abdomen would be poorly heated. It has been shown that at least parts of some large tumors, particularly eccentric tumors, are heatable by the Magnetrode and that some favorable responses have been observed when the treatment has been combined with chemotherapy or radiation therapy (Storm et al. 1981; Baker et al. 1982). There have been few systematic studies with good thermometry to document the heating patterns achieved. In reviewing the studies in which the thermometry was diligently approached, the heating patterns achieved are consistent with predictions from the physical energy deposition patterns predicted from theory (Oleson et al. 1983; Strohbehn 1984; Sapozink et al. 1985a). There is nothing to suggest that heat conduction or altered pathways of eddy currents by tumor interfaces improve the heating pattern over that predicted from the known physics of the device. In our experience, the complexities of body contour and bone–soft tissue interfaces in the sternum and sacrum universally lead to spots of severe discomfort for the patient which usually become treatment limiting. Often it has not been possible to document temperatures in the areas of pain, but overheating sites where electrical paths are compacted is the most logical explanation for this discomfort.

Capacitance heating has had some popularity for superficial tumor sites, but because of the selective power absorption in fat and the potential over-heating of subcutaneous tissue, interest has waned and there has been little activity in this country toward developing instrumentation for deep heating. In Japan, however, a device has been developed which operates at 8 MHz and preliminary results are sufficiently encouraging to reopen some interest in investigating this approach (Hiraoka et al. 1984; Kato et al. 1985). One suggestion has been to perform direct intercomparisons between the capacitance device and other devices for regional hyperthermia such as the annular phased array.

The largest experience reported using the annular phased array approach has been from the University of Utah where the instrument has been in use since 1980. The first report was a pilot study of the first 46 patients heated for massive abdominal or pelvic tumors (Sapozink et al. 1984). Most of the patients were treated with hyperthermia and radiation therapy with the latter treatment dependent to large measure on the extent and type of previous treatment. Though the patients were diverse, heavily treated in the past, and had very advanced disease with large tumor burdens, some important conclusions were drawn. It was possible to obtain heating in the pelvis and the abdomen, though this was often at considerable stress to the patient, either systemically for abdominal treatment or local pain and associated symptoms in the pelvis. Further, though a large number of usually relatively minor side effects were noted and occasional subacute effects, long-term complications were not a problem. Responses were not dramatic but many patients experienced subjective and objective improvement such that the results would be considered encouraging. Subsequently, a clinically useful method of calculation and expression of heating parameters in terms of 'thermal dose' was developed and published (Sapozink *et al.* 1985b). This approach was then applied to a direct intercomparison of the annular phased array and the concentric coil in patients treated successively on each device with identical or similar thermometry (Sapozink *et al.* 1985a). The conclusions of this study were that the heating patterns were more predictable with the concentric coil with heating in the outer annulus of the body and rapid drop off to zero or near zero centrally. Heating from the annular phased array was less predictable but often achieved effective heating centrally, particularly in the pelvis that was never seen with the concentric coil. In the abdomen there were insufficient intercomparisons to make a judgement of superiority of one device over the other, though neither seemed to be particularly effective; the concentric coil because of poor central power deposition and pain in the region of the sternum or the back being power limiting and the annular array because of systemic heating. More recently the experience with the annular array in pelvic tumors (Sapozink *et al.* 1986a) and abdominal tumors (Sapozink *et al.* 1986b) has been updated. In both locations an improved survival was noted in patients who had a response, either complete or partial. In the pelvis it was possible by logistic regression analysis to correlate response rates with various treatment and patient characteristic parameters. It was found that the dose of the concurrent radiation and the number of heat treatments at which a temperature of 42 °C was reached at some point during the treatment each independently correlated with response. It is clear from this experience that it is possible to achieve meaningful levels of hyperthermia in some patients, particularly in the pelvis, and that this is more difficult to achieve in the abdomen. The treatment is stressful to the patient and labor intensive on the part of the staff. As mentioned in the section on technical approaches, it is possible to achieve deep heating with ultrasound, either through techniques of moving transducers or arrays of focused transducers. To this point there have been few descriptions of clinical use published to review.

E. SUMMARY AND CONCLUSIONS

Clearly, hyperthermia has a strong and compelling biological rationale and this rationale is supported by the study of superficial hyperthermia in humans. There can now be no doubt that hyperthermia has efficacy against cancer in man. As more sophisticated approaches are developed, the biological rationale may be further extended, allowing one to capitalize on such phenomena as pH change, altered blood flow, thermotolerance, sequence events, altered pharmacokinetics, and likely many other subtleties of biology and metabolism. There are some areas where further preclinical experiments are needed to assist in more clearly defining the rationale for hyperthermia and chemotherapy. This is particularly true for local-regional heating methods with chemotherapy tested *in vivo*. Similarly, if whole-body hyperthermia is to have a role in clinical treatment, *in vivo* preclinical experiments to more firmly define a rationale would greatly enhance the attractiveness of experimental whole-body hyperthermia protocols in humans.

A major problem rests in the technical approaches to achieve effective clinical hyperthermia. One aspect of this problem is how to utilize better what equipment we have. What procedures can be used to minimize the patient's symptoms? Are there approaches that can make the treatments more efficient and less stressful to staff and patients? There remains a great deal to be done in terms of equipment development. In spite of the relative sophistication of the electronics of some of the available hyperthermic equipment, one need only to compare such equipment with a modern radiotherapy linear accelerator to realize that there is substantial room for technological improvement. In terms of ease of administration of the treatment, patient comfort, reproducibility of treatment, predictability of dose, predictability of efficacy and reliability of operation of the equipment, hyperthermia equipment ranks a distant second. Fundamental new approaches may be necessary to solve some of these problems. There should be more application of thermal dose to assessing treatment. This would, of course, be greatly facilitated if a noninvasive rapid turnaround system for thermometry were to be developed. And finally, in my view, the most significant and pressing research question in clinical hyperthermia is to define its role in the multidisciplinary treatment of cancer.

In spite of the difficulties in applying deep hyperthermia, I feel that the scientific and medical community should be optimistic and enthusiastic over the potential for hyperthermia in clinical cancer care. The biological rationale is extremely strong experimentally, both *in vitro* and *in vivo*. Of the various proposed modifiers of radiation therapy response, the rationale for hyperthermia is the strongest, has both additive and synergistic features, and has the added advantage of not relying on a single tumor characteristic such as the presence of hypoxic cells. Efficacy has been demonstrated in superficial human cancers. Even in the more difficult problem of deep heating, progress is being made and much has been learned in the past few years. The technological problems should be solvable with appropriate effort.

ACKNOWLEDGEMENTS
The author wishes to thank Frederic A. Gibbs, Jr., MD for helpful suggestions and critique, and Margaret D. Olsen for preparation of the manuscript.

Work at the University of Utah was supported by Contract Number NO1

CM 17523 and Grant Number PO1 CA 29578 awarded by the National Cancer Institute, Department of Health and Human Services.

REFERENCES

Aristizabal, S.A. and Oleson, J.R. (1984). Combined interstitial irradiation and localized current field hyperthermia: Results and conclusions from clinical studies. *Cancer Res. (Suppl.)* **44**, 4757–4760.

Baker, H.W., Snedecor, P.A., Goss, J.C., *et al.* (1982). Regional hyperthermia for cancer. *Am. J. Surg.* **143**, 586–590.

Ben Hur, E., Elkind, M.M. and Brock, B.V. (1974). Thermally enhanced radio-response of cultured Chinese hamster cells: Inhibition of repair of sublethal damage and enhancement of lethal damage. *Radiat. Res.* **58**, 34–51.

Coley, W. (1893). The treatment of malignant tumors by repeated inoculations of erysipelas: With a report of ten original cases. *Am. J. Med. Sci.* **105**, 487–511.

Coley, W. (1911). A report of recent cases of inoperable sarcoma successfully treated with mixed toxins of erysipelas and bacillus prodigiosus. *Surg. Gynecol. Obstet.* **13**, 174–190.

Coughlin, C.T., Douple, E.B., Strohbehn, J.W., Eaton, W.L., Trembly, B.S. and Wong, T.Z. (1983). Interstitial hyperthermia in combination with brachytherapy. *Radiology* **148**, 285–288.

Crile, G., Jr. (1962). Selective destruction of cancers after exposure to heat. *Ann. Surg.* **156**, 404–407.

Crile, G., Jr. (1963). The effects of heat and radiation on cancers implanted on the feet of mice. *Cancer Res.* **23**, 372–380.

Dewey, W.C., Hopwood, L.E., Sapareto, S.A. and Gerweck, L.F. (1977). Cellular responses to combinations of hyperthermia and radiation. *Radiology* **123**, 463–474.

Doyen, E. (1910). Traitement local des cancers accessibles par l'action de la chaleur au dessus de 55°. *Rev. de Therap. Med. -Chir.* **77**, 577.

Emami, B., Marks, J.E., Perez, C.A., Nussbaum, G.H., Leybovich, L. and von Gerichten, D. (1984). Interstitial thermoradiotherapy in the treatment of recurrent/residual malignant tumors. *Am. J. Clin. Oncol.* **7**, 699–704.

Gerweck, L. (1977). Modifications of cell lethality at elevated temperatures: The pH effect. *Radiat. Res.* **70**, 224–235.

Gerweck, L.E., Dahlberg, W.K., Epstein, L.F. and Shimm, D.S. (1984). Influence of nutrient and energy deprivation on cellular response to single and fractionated heat treatments. *Radiat. Res.* **99**, 573–581.

Gibbs, F.A., Jr. (1983). 'Thermal mapping' in experimental cancer treatment with hyperthermia: Description and use of a semi-automatic system. *Int. J. Radiat. Oncol. Biol. Phys.* **9**, 1057–1063.

Gibbs, F.A., Jr. and Stewart, J.R. (1985). Regional hyperthermia in the treatment of cancer: A review. *Cancer Invest.* **3**, 445–452.

Hahn, G.M. (1979). Potential for therapy with drugs and hyperthermia. *Cancer Res.* **39**, 2264–2268.

Henle, K.J. (1981). Interaction of mono- and polyhydroxy alcohols with hyperthermia in CHO cells. *Radiat. Res.* **88**, 392–402.

Henle, K.J. and Dethlefsen, L.A. (1980). Time-temperature relationships for heat-induced killing of mammalian cells. *Ann. N.Y. Acad. Sci.* **335**, 234–253.

Hiraoka, M., Jo, S., Takahashi, M. and Abe, M. (1984). Treatment of locally advanced hepatocellular carcinoma with RF hyperthermia. In *Hyperthermic Oncology 1984 – Proc. of the 4th Internat. Symp. on Hyperthermic Oncology, Vol. 1*, pp. 803–806. Overgaard, J. (Ed.), Taylor and Francis, London and Philadelphia.

Kato, H., Hiraoka, M., Nakajima, T. and Ishida, T. (1985). Deep-heating characteristics of an RF capacitive heating device. *Int. J. Hyperthermia* **1**, 15–28.

Lele, P.P. and Parker, K.J. (1982). Temperature distribution in tissues during local hyperthermia by stationary or steered beams of unfocused or focused ultrasound. *Br. J. Cancer* **45**, 108–121.

Li, G.C., Shiu, E.C. and Hahn, G.M. (1980). Similarities in cellular inactivation by hyperthermia or by ethanol. *Radiat. Res.* **82**, 257–268.

Marmor, J. (1983). Cancer therapy by ultrasound. *Adv. Radiat. Biol.* **10**, 105–133.

Meyer, J.L. (1984). The clinical efficacy of localized hyperthermia. *Cancer Res. (Suppl.)* **44**, 4745–4751.

Oleson, J.R. (1982). Hyperthermia by magnetic induction: I. Physical characteristics of the technique. *Int. J. Radiat. Oncol. Biol. Phys.* **8**, 1747–1756.

Oleson, J.R., Heusinkveld, R.S. and Manning M.R. (1983). Hyperthermia by magnetic induction: II. Clinical experience with concentric electrodes. *Int. J. Radiat. Oncol. Biol. Phys.* **9**, 549–556.

Paliwal, B.R., Gibbs, F.A., Jr. and Wiley, A.L. (1982). Heating patterns induced by a 13.56 MHz radiofrequency generator in large phantoms and pig abdomen and thorax. *Int. J. Radiat. Oncol. Biol. Phys.* **8**, 857–864.

Sapareto, S.A. and Dewey, W.C. (1984). Thermal dose determination in cancer therapy. *Int. J. Radiat. Oncol. Biol. Phys.* **10**, 787–800.

Sapozink, M.D., Gibbs, F.A., Jr., Gates, K.S. and Stewart, J.R. (1984). Regional hyperthermia in the treatment of clinically advanced deep-seated malignancy: Results of a study employing an annular array applicator. *Int. J. Radiat. Oncol. Biol. Phys.* **10**, 775–786.

Sapozink, M.D., Gibbs, F.A., Jr., Thomson, J.W., Eltringham, J.R. and Stewart, J.R. (1985a). A comparison of deep regional hyperthermia from an annular array and a concentric coil in the same patients. *Int. J. Radiat. Oncol. Biol. Phys.* **11**, 179–190.

Sapozink, M.D., Gibbs, F.A., Jr., and Sandhu, T.S. (1985b). Practical thermal dosimetry. *Int. J. Radiat. Oncol. Biol. Phys.* **11**, 555–560.

Sapozink, M.D., Gibbs, F.A., Jr., Egger, M.J. and Stewart, J.R. (1986a). Regional hyperthermia for clinically advanced deep-seated pelvic malignancy. *Am. J. Clin. Oncol.* In press.

Sapozink, M.D., Gibbs, F.A., Jr., Egger, M.J. and Stewart, J.R. (1986b). Abdominal regional hyperthermia with an annular phased array. *J. Clin. Oncol.* **4**, 775–783.

Scott, R.S., Johnson, R.J.R., Story, K.V. and Clay, L. (1984). Local hyperthermia in combination with definitive radiotherapy: Increased tumor clearance, reduced recurrence rate in extended follow-up. *Int. J. Radiat. Oncol. Biol. Phys.* **10**, 2119–2123.

Song, C.W., Rhee, J.G. and Levitt, S.H. (1982). Effect of hyperthermia on hypoxic cell fraction in tumor. *Int. J. Radiat. Biol.* **8**, 851–856.

Stewart, J.R., Gibbs, F.A., Lehman, C.M., Peck, J.W. and Egger, M.J. (1983). Change in the *in vivo* hyperthermic response resulting from the metabollic effects of temporary vascular occlusion. *Int. J. Radiat. Oncol. Biol. Phys.* **9**, 197–201.

Storm, F.K., Elliott, R.S., Harrison, W.H., Kaiser, L.R. and Morton, D.L. (1981). Radio-frequency hyperthermia of advanced human sarcomas. *J. Surg. Oncol.* **17**, 91–98.

Strohbehn, J.W. (1982). Theoretical temperature distributions for solenoidal-type hyperthermia systems. *Med. Phys.* **9**, 673–682.

Strohbehn, J.W. (1984). Calculation of absorbed power in tissues for various hyperthermia devices. *Cancer Res. (Suppl.)* **44**, 4781s–4787s.

Turner, P.F. (1984). Regional hyperthermia with an annular phased array. *IEEE Trans. Biomed. Eng.* **31**, 106–114.

Urano, M. and Kahn, J. (1983). The change in hypoxic and chronically hypoxic cell fraction in murine tumors treated with hyperthermia. *Radiat. Res.* **96**, 549–559.

Wallen, C.A., Colby, T.V. and Stewart, J.R. (1986). Cell kill and tumor control after heat treatment with and without vascular occlusion in RIF-1 tumors. *Radiat. Res.* **106**, 215–223.

Westra, A. and Dewey, W. (1971). Variation in sensitivity to heat shock during the cell cycle of Chinese hamster cells *in vitro*. *Int. J. Radiat. Biol.* **19**, 467–477.

Yau, T.M. (1979). Procaine-mediated modification of membranes and of the response to X-irradiation and hyperthermia in mammalian cells. *Radiat. Res.* **80**, 523–541.

Hyperthermia and Oncology, Vol. 1, pp. 13–56 (1988)
Urano and Douple (Eds.)
© 1988 VSP.

Chapter 2

The effects of hyperthermia on cellular macromolecules

JOSEPH L. ROTI ROTI and ANDREI LASZLO
Washington University School of Medicine, Mallinckrodt Institute of Radiology, Division of Radiation Oncology, Section of Cancer Biology, St. Louis, MO 63108, USA

A. INTRODUCTION

One of the properties of hyperthermia which makes its use in cancer therapy feasible is the ability of heat to systematically kill cells. The kinetics of this killing process has been established in the late 1970s (reviewed by Conner *et al.* 1977; Henle and Dethlefsen 1978; Dewey *et al.* 1980). However, the nature of the critical subcellular target(s) and the molecular mechanism(s) of heat-induced cell killing remain uncertain in the face of continued research efforts (Leeper 1985; Streffer 1985). Nevertheless, it seems reasonable to assume that cells are killed as a result of thermal damage to one or more subcellular and macromolecular systems.

In a lethal event due to hyperthermia 10^3 to 10^5-fold more energy is expended than that expended by ionizing radiation (Hall 1978; Hahn 1982). Clearly, such large amounts of energy input cause multiple and pleiotropic effects which confound elucidation of the mechanisms leading to cell death. The situation is further complicated by the wide variety of cell types and heating conditions employed by the different investigators in this field. These complications make the elucidation of the mechanisms for cell killing a formidable undertaking. To this end, it would be ideal to make comparisons of heat effects on all cellular systems at iso-survival levels, but we have not found this possible. In order to include all relevant studies we have defined the experimental conditions and systems used. If data were collected over a broad range of heat exposures, the conclusions of such work will be given in general statements.

The purpose of this chapter is to review the effects of heat on macromolecular systems with the goal of generating a set of working hypotheses regarding the mechanisms of cell killing. Clearly, such hypotheses are seen not as an end in themselves, but as steps in the process to define more precisely the lethal effects of heat on cells. Thus, the overall approach in this review is to examine heat effects on macromolecules in the context of the organelle systems in which they exist. We begin by describing the plasma membrane, proceed through the cytoskeleton and cytosol to the nucleus. It is hoped that this organization will assist the reader in integrating the many diverse effects of hyperthermia.

B. MEMBRANES

The plasma membrane is the cellular component which is in immediate contact with the environment. Thus, it is expected that the plasma membrane is affected by hyperthermia in a variety of ways. In fact, it appears that hyperthermia-induced alterations in the plasma membrane play a major, if not primary, role in the cellular effects of hyperthermia. Agents that are known to act specifically at the membrane, including local anesthetics and aliphatic alcohols, all act synergistically with heat (Yatvin 1977; Li and Hahn 1978). The cellular inactivation due to the action of these agents by themselves is strikingly similar to the action of the heat alone (Li et al. 1980). Overall, the studies summarized above indicate that alterations in the plasma membrane must be involved in cellular responses to hyperthermia. The nature of changes in cell membrane structure exposed to hyperthermia has been investigated in some detail.

1. Membrane lipids

The current view of membrane structure indicates that the plasma membrane is composed primarily of an asymmetric lipid bilayer which also contains integral proteins (Alberts et al. 1983). The lipids involved in membranes are phospholipids. The hydrophilic polar heads of these lipids, under the cellular ionic conditions, are oriented toward the outer surfaces of the bilayer while the hydrocarbon tails are found in the interior. The degree of saturation of carbon–carbon bonds in the hydrocarbon chain determines the order of the overall structure. The saturated hydrocarbon phospholipids are restricted in their molecular motions, as determined by techniques such as electron spin resonance (ESR) or fluorescence polarization (Alberts et al. 1983). The degree of restriction of molecular motion is defined as 'order'. The unsaturated hydrocarbon side chains allow more molecular motion; this additional freedom of motion or loss of 'order', is defined as 'fluidity'. Pure lipid bilayers are thought to undergo phase transitions at different temperatures, resulting in different molecular orders. Below transition temperatures, the lipids are in a solid-like state, while at temperatures above the phase transitions, the lipid bilayer acquires the properties of the fluid state. However, such transitions are masked in membranes of mammalian cells by the presence of proteins and cholesterol. Cholesterol appears to be involved in the regulation of fluidity, while the presence of proteins in the plasma membrane leads to nonuniformity of viscosity. The relative properties of membrane microregions depend both on the exact composition of the membrane, i.e. the levels of cholesterol, the ratio of saturated to unsaturated phospholipids, etc. (Alberts et al. 1983) and on the physiology of the cell (Lai et al. 1980).

Lipid composition and thermosensitivity. Numerous studies suggest a relationship between thermal sensitivity and the lipid composition of the membranes.

Composition. Varying the growth temperature has been found to alter the lipid composition of bacterial membranes (Yatvin 1977) to maintain a relatively constant state of fluidity. These changes were accompanied by changes in survival after hyperthermic treatment, showing that membrane composition may be related to hyperthermic sensitivity.

Similarly, the membrane composition of mammalian cells has been altered experimentally, both *in vivo* and in culture. Murine P388 cells grown in animals fed a diet high in polyunsaturated fatty acids were shown to be more thermosensitive than cells grown in animals fed a diet high in saturated fatty acids (Hidvegi *et al*. 1980; Mulcahy *et al*. 1981). Growing L1210 leukemia cells in media supplemented with highly polyunsaturated fatty acids led to increased thermosensitivity, while decreased thermosensitivity was observed when cells were supplemented with more saturated fatty acids (Guffy *et al*. 1982). Similar results were reported recently with mouse fibroblast LM cells grown in media containing various fatty acids (Konings 1985; Konings and Ruifrock 1985).

There have been reports indicating that the cholesterol content and cholesterol/phospholipid ratios of several cultured cell lines correlates inversely with their thermosensitivity (Cress and Gerner 1980; Cress *et al*. 1982). However, recent results obtained with thermotolerant cells and heat resistant variants of melanoma cells have challenged the universality of these conclusions (Gonzalez-Mendez *et al*. 1982; Anderson *et al*. 1984). Growth of CHO cells at temperatures higher than usual, 39 and 41 °C, results in increased heat resistance (Li and Hahn 1980). When such heat adapted cells reach confluence, the heat sensitivity did correlate with cholesterol phospholipid ratios and membrane fluidity, but these correlations did not extend to cultures growing exponentially at these temperatures (Anderson *et al*. 1981).

Role of lipids in cell killing. Studies with a lipid soluble fluorescent probe of plasma membranes isolated from V79 cells which had been heated at various temperatures indicated two phase transitions involving lipids, one at 8 °C, and a broader transition between 23 ° and 36 °C, centered at about 30 °C (Lepock *et al*. 1981; Lepock 1982). There was no evidence for a lipid phase transition at the temperature associated with the onset of cell killing e.g., 41.5 °C. Similar results were obtained using an ESR spin label probe, 2N14 (Lepock *et al*. 1983). In another study using mouse lymphoma cells, the diffusion coefficient of the lipid probe, DII, indicated that no perturbation of lipid fluidity occurred at 41–45 °C (Mehdi *et al*. 1984). Thus, it appears unlikely that effects on the lipid components of the plasma membrane are responsible for heat-induced cell killing.

2. Membrane proteins
Many altered plasma membrane functions induced by hyperthermic treatments are those which are presumbly mediated in part, if not entirely, by the protein

components of the membrane. The presence of phase transitions in the protein phase of the membrane at hyperthermic temperatures was investigated in V79 cells by measuring intrinsic protein fluorescence and energy transfer from membrane protein to the lipid soluble fluorescent probe, trans-paranaric acid (Lepock *et al.* 1983). These measurements indicated that irreversible protein transitions in membranes occurred at temperatures of 40 ° to 41 °C. Such results would argue that hyperthermic killing may be associated with the effects on the protein component of the plasma membrane.

Receptor proteins. The binding of epidermal growth factor to Rat-1 fibroblasts was inhibited by hyperthermia at 45 °C (Magun and Fennie 1981). Scatchard analysis indicated that this inhibition was due to a decreased affinity of the receptors for the ligand. Exposure of Chinese hamster HA-1 cells to hyperthermia at 43 ° to 45 °C led to a time–temperature dependent inhibition of insulin binding to its receptor (Calderwood and Hahn 1983). However, Scatchard analysis of the binding data revealed that, in this case, heat acted by reducing receptor number rather than altering the affinity for the hormone. It is possible that such loss of activity could be due to the extensive blebbing associated with such hyperthermic treatments (Coss *et al.* 1982; Borrelli *et al.* 1986). Hyperthermia (43 ° to 45 °C) inhibits the binding of monoclonal antibodies to histocompatibility antigens on the surface of murine lymphoma cells in suspension culture. This inhibition also appears to result from a reduction of receptor number (Mehdi *et al.* 1984). These studies demonstrate that hyperthermic temperatures affect membrane proteins.

Membrane transport proteins. The plasma membrane acts to establish gradients of low molecular weight solutes and ions between the cell and its environment. This situation is accomplished by a combination of passive diffusion and the action of ion specific pumps and solute specific transporter systems. The functioning of these systems involves membrane proteins (Alberts *et al.* 1983).

Hyperthermia alters the membrane permeability to several compounds, including adriamycin, (Hahn and Strande 1976), polyamines (Gerner *et al.* 1980) and certain ions (Strom *et al.* 1977). The membrane also becomes more 'leaky' also with respect to the inside of the cell. Heated cells have been found to release more uridine and inorganic phosphate (Hayat *et al.* 1984). In fact it was reported that heat resistant variants of pig kidney cells released less uridine after heating than their normal counterparts (Reeves 1972). Heated cells have been observed to release K^+ ions (Ruifrock *et al.* 1985a, b). Some of this release may be a consequence of the extensive blebbing mentioned earlier and to loss of vehicles containing Na^+/K^+ ATPase, as observed by these workers.

Hyperthermia appears to alter the transport functions of the plasma membrane. Hyperthermia at 43.5 ° and 45 °C inhibited uptake of thymidine in CHO cells grown in suspension and in monolayers (Slusser *et al.* 1982). Exposure of rat and human thymocytes to temperatures ranging from 39 ° to 43 °C resulted in striking inhibition of Na^α dependent amino acid transport

(Kwock *et al.* 1978; Lin *et al.* 1978). Uridine uptake was inhibited in pig kidney cells treated at 46 °C (Reeves 1972) and Hoechst 33342 uptake was inhibited in CHO cells treated at 45.5 °C (Rice *et al.* 1985b). Leucine and uridine uptake decreased in rat-1 fibroblasts exposed to 45 °C for 30 minutes (Magun 1981). Exposure of HeLa cells in suspension to 42 °C for 1 hour inhibited uridine uptake slightly and bleomycin uptake dramatically (Souliman and Chapman 1981). Amino acid and uridine uptake were not inhibited by exposure of CHO cells to 45 °C for 10 and 17.5 minutes (Henle and Leeper 1979); however, in this study the earliest time point examined was 3 hours after heating which was longer compared to the previously cited studies. Overall, the studies summarized above indicate that exposure to hyperthermia inhibits the transport of small metabolites; at the heat doses studied, such inhibition appears reversible, with the possible exception of the thymocytes.

The status of ion influx and efflux post-hyperthermia is equivocal. Treatment of CHO cells at 42 °C led to an increased influx of K^+ ions within 15 minutes (Stevenson *et al.* 1983), which was reversed immediately upon return to 37 °C. There was a slight decrease in the total K^+ content after 150 minutes at 42 °C, indicating that efflux rates also must have been changed. When mouse mastocytoma P815 cells grown in suspension were exposed to elevated temperatures, significant decreases in total K^+ and Cl^- content were observed. A 40 per cent decrease in K^+ was observed after 43 °C for 60 minutes (Yi 1979, 1983). Similar results were obtained with mouse LM fibroblasts grown in suspension treated at 44 °C; K^+ content decreased in a dose dependent manner, leveling off at 16 hours after treatment (Ruifrock *et al.* 1985b). However, when Reuber hepatoma H3T in monolayers were exposed to 42 °C for 30 minutes, no changes in the steady state levels of Na^+ or K^+ were observed (Boonstra *et al.* 1984). Similarly, exposing plateau-phase cells in monolayers to 45 °C for up to 30 minutes had no significant effect on the Na^+, K^+, Mg^{2+} content by 28 hours post-heat, but did lead to a dose dependent increase in Ca^{2+} content by 22 hours post-heat (Vidar and Dewey 1986). Microelectrode measurements of individual neuroblastoma cells exposed to 45.5 °C for 40 minutes indicated no changes in free intracellular K^+ and Cl^- concentrations immediately after heating and for post-heating times of up to 30 hours (Borrelli *et al.* 1986b). The results in the last three studies were obtained using cells in monolayer, thereby excluding the cells physiologically dead immediately or soon after hyperthermia. In contrast, the earlier studies used cells grown in suspension, which included such dead cells. Thus it is likely that the data showing alterations in ion content due to hyperthermia were obtained from heterogeneous live–dead cell populations.

Overall we are left with a paradox. The biophysical data strongly suggest that protein alterations are involved in the response of plasma membrane to hyperthermia. However, when specific functions are examined, it is striking that the plasma membrane is very resilient to heat, maintaining function under conditions that lead to greater than 98 per cent cell killing, in some cases. This fact may indicate that the cell's metabolic integrity is much more resistant to heat than its reproductive capacity.

A late developing membrane effect. In a recent study, hyperthermia-treated mammalian cells were shown to be differentially stained with the fluorescent probe *N*-Σ-dansyl-L-lysine 12 to 24 hour post-heat (Rice *et al.* 1985a). This probe has been shown to have higher solubility in phosphatidylcholine liposomes of low cholesterol content than in those of high cholesterol content (Humphries and Lovejoy 1983). Preliminary evidence suggests that the *N*-Σ-dansyl-L-lysine staining regions are cholesterol-free domains (Rice *et al.* 1985a). *N*-Σ-dansyl-L-lysine staining has been found to correlate well with cell killing (Rice *et al.* 1985a) and this method may be useful for separating live and dead cells. Biochemical analysis of clonogenically dead versus live cells could lead to an understanding of which heat-induced membrane alterations play a role, if any, in heat-induced cell killing.

3. Summary of membrane effects
The above results clearly indicate that hyperthermia has a significant effect on the structure and function of the plasma membrane, in spite of its metabolic resiliency. It is possible that some of these effects are primary, while others may be secondary in the induction of reproductive death. The immediate consequences of these membrane effects could play a triggering role in the constellation of other pleiotropic cellular changes induced by hyperthermia, such as cytoskeletal and nuclear alterations.

C. CYTOSKELETON
The typical mammalian cells used in the studies of the effect of elevated temperatures contain cytoskeletal filamentous networks composed of three types of filaments: microfilaments (MF), microtubules (MT) and intermediate filaments (IF) (Alberts *et al.* 1984). Alterations of the organization of all three major cytoskeletal components have been reported after exposure of cells to elevated temperatures. The nature of these changes depends, however, on the temperature of the treatments and the type of cells studied.

1. Microfilaments
In cultured cells the actin containing MF of 60 Å diameter forms cytoplasmic bundles that are called 'stress filaments'. Stress filaments can be observed by light microscopy using indirect immunofluorescence with actin antibodies or a fluorescently labeled actin specific metabolite, phalloidin. Stress filaments have been shown to contain myosin, α actinin and tropomyosin in addition to actin. Only the actin components have been studied currently in terms of the effects of hyperthermia.

Exposure of HeLa or gerbil fibroma cells to 42 °C for several hours did not result in any significant change in stress filament organization; on the contrary, the cells became flatter and the number of stress filaments increased (Thomas *et al.* 1982). Similar observations were made in rat embryo fibroblast exposed to 42 °C for 3 hours (Welch and Suhan 1985). However, in the later study, actin-containing bundles were observed in the nucleus. This result implies that cytoplasmic elements translocated into the nucleus. On the other hand, ultrastructural observations of CHO cells heated in mitosis indicated that, both a brief acute treatment (45.5 °C for 6 minutes) and a prolonged mild treatment (41.5 °C for 7 hours) led to the disruption of MF. The former treatment disrupted the actin-containing contractile ring, while the latter destroyed the cortical MF (Coss *et al.* 1979, 1982).

A differential response of stress filaments in Reuber H35 hepatoma cells and in neuroblastoma N2 cells to a 43 °C, 30-minutes treatment was recently reported (Van Bergen En Henegousen *et al.* 1985). The treatment resulted in the disruption of stress filaments in the hepatoma cells, but not in the neuroblastoma cells. When the heat exposure was varied between 40 ° and 45 °C for 30 minutes the difference in the response of the stress filaments of the two cell types remained unaltered, showing that it was not due to a threshold effect.

Exposure of interphase CHO cells to 45 °C led to a rapid disruption of stress fibers. Within 5 minutes of treatment, 90 per cent of the cells did not contain observable stress fibers (Glass *et al.* 1985). These changes were reversible: intact stress fibers were observed in greater than 90 per cent of the cells by 24 hours after treatment. There appears to be no direct correlation between the heat-induced loss of stress fibers in CHO cells and heat-induced cell death.

The mechanism of stress fiber loss could be depolymerization of the F-actin containing MF, dissociation of the MF bundles, or a combination of both. There have been few studies concerning the effect of hyperthermia on actin *in vitro*. One study indicated that G-actin was inactivated when exposed to temperatures from 42 ° to 55 °C, as determined by its ability to inhibit DNAse I activity (Heacock *et al.* 1982).

One difficulty with the cellular studies outlined above is the fact that hyperthermia alters cell shape and morphology, which may be related to stress fiber organization; thus the causal links are difficult to ascertain. Moreover, the molecular mechanism(s) of stress fiber formation are obscure, although there exist factors that promote the polymerization of G-actin to F-actin (Schliwa 1981; Korn 1982) and an integral membrane protein has been isolated which promotes the 'bundling' of MF into stress fibers (Brown *et al.* 1983). Cultured cells must attach to the substrate in order to form stress fibers. The formation of stress fibers appears to require transmembrane linkages. Since hyperthermia has a pleiotropic effect on the plasma membrane (reviewed above), the loss of the stress fibers might be a secondary effect as a reflection of these changes. In addition, the changes in ion fluxes induced by hyperthermia may alter factors involved in stress filament formation. More information concerning the effects of hyperthermia on stress fiber attachment points and assembly factors are needed to ascertain the molecular mechanisms involved.

2. *Microtubules*
The MT play a key role in the orderly segregation of the genetic material at cell division by forming the mitotic spindle. Obviously, disruption of the mitotic apparatus could lead to death of the heat sensitive mitotic cells. In

interphase cells, MT are organized into a delicate network in the cytoplasm; the exact function of this network is not clear at this time.

Spindle microtubules. When CHO cells were exposed to 41.5 °C for several hours (up to 7), the cells entered mitosis at one-half the normal rate, and displayed a much lengthened metaphase (Coss *et al.* 1979). Since the cells divided eventually, the mitotic apparatus must be functioning under these conditions. Indeed, ultrastructural examination of these cells indicated that both pole to pole and pole to kinetochore MT were formed under these conditions. However, exposure of mitotic CHO cells to 45.5 °C led to the destruction of the mitotic spindle. After 6 minutes, the spindle was disorganized, and by 15 minutes of heating there was a complete disassembly of the spindle (Coss *et al.* 1982).

Cytoplasmic microtubules. The effect of hyperthermia on the organization of cytoplasmic MT has also been examined. Exposure of HeLa cells, gerbil fibroma cells, and rat embryonic fibroblasts to 42 °C did not disturb the cytoplasmic MT network (Thomas *et al.* 1982; Welch and Suhan 1985). Similarly, hyperthermic treatments (30 minutes at 40 ° to 45 °C) did not alter interphase MT organization in Reuber hepatoma H35 cells (Van Bergen En Henegousen 1985). However, the cytoplasmic MT network was destroyed in 3T3 mouse fibroblasts exposed to 43 °C for 30 minutes (Lin *et al.* 1982) and in N2A neuroblastoma cells exposed to 40 ° to 45 °C for 30 minutes (Van Bergen En Henegousen 1985).

Microtubule polymerization. MT are polymerization products of monomers of the subunit protein, tubulin. Although tubulin polymerization has been well characterized, there have been few studies on the effect of hyperthermia on the *in vitro* assembly of tubulin. In crude brain extracts, shifting the temperature from 37 ° to 41 °C led to a reversible depolymerization of tubulin (Lin *et al.* 1982). The addition of glycerol gave a 2 to 3-fold protection from this effect. A more purified tubulin, i.e. cycled twice through polymerization–depolymerization, was not sensitive to treatment at 41 °C, but became sensitive upon the addition of the supernatant from the crude extract devoid of tubulin or by exposures to 43 °C (Lin *et al.* 1982). Depolymerization under these conditions was not reversible. In a more comprehensive study, tubulin cycled 3.5 times was used (Coss *et al.* 1982). Heating (up to 50 °C) tubulin under conditions when assembly was biochemically blocked caused subsequent assembly to occur at a depressed rate and to a lesser extent. Incubation of already assembled MT at temperatures up to 50 °C followed by deassembly–reassembly led to similar results, except that more protein appeared to be inactivated under these conditions. Interphase MT exposed to hyperthermia are disassembled, but can be reassembled albeit at slower rates and to a lesser extent. However, the mitotic spindle cannot reform in mitotic cells heated at 45.5 °C and then incubated at 37 °C (Coss *et al.* 1982). This observation suggests that the ability of centrosomes to nucleate MT was impaired by the 45.5 °C treatment. The disrupted morphology of centrosomes observed under these conditions supports this hypothesis (Barrau *et al.* 1978; Coss *et al.* 1982).

Thus, the heat-induced changes in MT organization in intact cells appears to be the result of both intrinsic denaturation of tubulin's capacity for assembly, and possible destruction or inhibition of the function of the centrosomes, which are the organelles associated with the *in vivo* control of assembly.

3. Intermediate filaments

The IF are the third component found in the cytoskeleton of mammalian cells. The type of IF varies according to the developmental origin of the cell. In spite of much effort, the function of IF is obscure at this time.

The effect of hyperthermia on IF organization has been examined by immunofluorescence and electron microscopy. Vimentin-containing IF collapsed after exposure to 42 °C, forming a perinuclear cap rich in IF in HeLa cells, gerbil fibroma cells, rat fibroblasts and hamster cells (Falkner *et al.* 1981; Thomas *et al.* 1982; Welch and Suhan 1985). Incubation of cells under normal conditions after hyperthermia led to eventual restoration of the normal distribution of IF. The time course and requirements for macromolecular synthesis of this recovery process have yet to be examined. Exposure to hyperthermia (40 ° to 45 °C) leads to disruption of vimentin organization in N2A neuroblastoma cells, while leaving both vimentin and cytokeratin organization intact in Reuber H35 hepatoma cells (Van Bergen En Henegousen *et al.* 1985). No studies of the effect of hyperthermia on IF assembly *in vitro* have been performed so far.

4. Summary

Hyperthermia leads to alterations in the organization of various components of the cytoskeleton; the severity and type of alterations depend on the temperature and length of treatment and on the cell type examined. The evidence so far indicates that heat (1), induces direct alterations in cytoskeletal components, interfering with their ability to self-assemble; and (2), modifies global mechanisms involved in the overall *in vivo* control of the assembly of the cytoskeleton. The exact mechanism of heat-induced alterations in cytoskeletal organization are yet to be elucidated. Except for cells obtained by mitotic shakeoff, all the above studies have been performed on cells grown in monolayers. Studies on the effect of hyperthermia on the organization of the cytoskeleton in cells heated in suspension are needed. It is reasonable to postulate that the cytoskeleton may provide a structural continuity between the plasma membrane and the nucleus. Hyperthermia induced changes in at least one cytoskeletal element in all cell types studied so far. Therefore, it is tempting to postulate that hyperthermia-induced disruption of the cytoskeleton plays a role in the heat-induced increase in nuclear protein content. This concept forms one of the main premises of our working model for heat-induced cell killing (see Overview).

D. CYTOSOL

1. Changes in morphology

Numerous heat-induced changes in cytoplasmic organization have been reported. Heine *et al.* (1971) examined HeLa cells grown in suspension, exposed to temperatures from 41 ° to 45 °C. The mitochondria, the endoplasmic reticulum and the Golgi apparatus did not exhibit any major changes in the number or form following hyperthermia (43 °C, 30 minutes). However, the lysosome number was greater in heated than in control cells at 24 hours posthyperthermia. Moreover, polysomes were not observed in heated cells, coincidental with the inhibition of protein synthesis (see Section C). However, in rat fibroblasts, heat treatment (42 °C, 3 hours) led to a fragmentation and/or disappearance of the endoplasmic reticulum and the Golgi complex (Welch and Suhan 1985). This change was accompanied by an increase in the number of vesicularized membranes within the perinuclear region.

a. Mitochondria. A number of structural changes occurred within the mitochondria in rat fibroblasts exposed to 42 °C for 3 hours; they appeared swollen, the cristae were more prominent, and the intracisternal spaces appeared enlarged (Welch and Suhan 1985). Small phase-dense structures observed around mitochondria of control cells, thought to be free ribosomes and/or polysomes, were no longer evident around the mitochondria of heated cells. This result may have been due to changes in the intracellular distribution of the mitochondria after heat shock. An increase in the number of the mitochondria located near and around the nucleus with fewer near the cell periphery was observed. This relocalization may be due to a collapse and aggregation of the intermediate filaments around the nucleus in the heated cells (see Cytoskeleton). Similar alterations in the structure of the mitochondria have been noted in chick myoblasts exposed to various uncouplers of oxidative phosphorylation (Buffa *et al.* 1970). In CHO cells grown in monolayer, exposure to 41.5 °C also caused the appearance of swollen mitochondria (Coss *et al.* 1979). The relationship between heat-induced morphological changes and possible alterations of mitochondrial function remains to be determined.

b. Lysosomes. Lysosomes are membranous organelles located in the cytoplasm containing hydrolytic enzymes, including proteases, nucleases, glycosidases, lipases, phospholipases, phosphatases, and sulfatases, which are used for the controlled intracellular digestion of macromolecules (Alberts *et al.* 1983).

An increase in the number of lysosomes was observed in HeLa cells 24 hours after a 43 °C, 2-hour treatment (Heine *et al.* 1971). A destruction of lysosomes was observed a few hours after treatment in solid mouse mammary tumors heated *in vivo* at 42.5 °C for 30 minutes (Overgaard 1976). Lysosomes isolated from tumor cells were found to be more sensitive to heat than their normal counterparts (Turano *et al.* 1970). Increased lysosomal enzyme activity was found in heated mouse mammary tumors (Overgaard and Overgaard 1972a, b), indicating that the heat-induced structural alterations led to functional alterations. Increased acid phosphatase activity was observed in the spleen of mice exposed to hyperthermia (Hume and Field 1977), indicating that heat-induced lysosomal alterations are not restricted to tumor cells. This effect was different according to the severity of the treatment (Hume *et al.* 1978). Between 41.0 ° and 42.3 °C, the enzyme activity reached a maximum 1.5 hours after heating and then decayed. At temperatures above 42.5 °C, enzyme activity increased immediately after heating and the levels of enzyme activity stayed high for at least 4 hours. The heat-induced alterations in lysosomal structure and function probably occur as a consequence of damage to lysosomal membranes. Since the effects observed occur immediately or soon after heating, they could play a major role in the interphase hyperthermic death of cells. It is difficult to envision at this time what role these alterations could play in hyperthermic reproductive death.

2. Respiration and glycolysis

Two of the major functions associated with the cytoplasm are energy metabolism via respiration and glycolysis. Inhibition of respiration by hyperthermia in tumors has been known for a long time (Westermark 1927). Exposure of Novikoff and 'minimal deviation 5123 hepatoma' cells to 43 °C led to an inhibition of oxygen uptake and of aerobic glycolysis (Mondovi *et al.* 1969). The effect was not observed, however, if cells were disrupted before exposure to elevated temperatures. Their studies with isolated mitochondria were inconclusive. In contrast, exposure of mitochondria isolated from mouse brain, liver and Ehrlich ascites tumor cells to 37 ° to 45 °C for 10 minutes led to an inhibition of electron transport, a loss of respiratory control and an uncoupling of oxidative phosphorylation (Christiansen and Kvamme 1969). The brain mitochondria were most heat resistant while ascites mitochondria were most heat sensitive. Oxidative phosphorylation of mitochondria prepared from heat-treated ascites cells was more heat resistant than that of mitochondria heated *in vitro*. Metabolic studies of sarcomas heated *in situ* were performed with tumor slices prepared after the treatment and incubated at 38 °C (Dickson and Calderwood 1979). Heating of the tumor for 1 hour at 40 °C had no influence on subsequent respiration or anaerobic glycolysis. However, exposure to 42 °C for 1 hour led to a 50 per cent inhibition of respiration, a 75 per cent inhibition of anaerobic glycolysis and a 90 per cent inhibition of aerobic glycolysis. These studies demonstrate that hyperthermic exposure inhibits metabolic functions involved in energy production. The exact mechanisms involved and whether they are associated with structural changes in the mitochondria remains to be determined.

3. Protein synthesis

One important cellular function associated with the cytosol is protein synthesis. This process is very heat sensitive, with many components of the process being affected by heat. For example, in CHO cells, exposure to 43 °C was found to inhibit both the elongation and initiation phases of protein synthesis (Oleinick 1979).

a. Overall effects on protein synthesis. Effects of hyperthermia on protein synthesis have been investigated in L5178Y murine leukemia lymphoblasts (Fuhr 1974; Fuhr and Overton 1974; Fuhr *et al.* 1974). The rate of protein synthesis was inhibited proportionately to the increase in temperature (up to 44 °C), and the recovery from the inhibition was slower in cells exposed to higher temperatures. In HeLa cells grown in suspension, incubation at 42 °C results in 4-fold decrease in protein synthesis within 10 minutes. This inhibition is reversible if cells are returned to 37 °C (McCormick and Penman 1969). Incubation of CHO cells at 45 °C for 10 minutes reduced protein synthesis to 3 per cent of control, while a 17.5-minutes treatment resulted in total inhibition (Henle and Leeper 1979). Recovery began approximately 4 hours after treatment. Recovery from 10 minutes at 45 °C was completed by 8 h after the initial treatment, while 26 hours were required in cells exposed to 45 °C for 17.5 minutes (Henle and Leeper 1979).

The mechanism of heat-induced inhibition of protein synthesis and recovery from it have been investigated using several different approaches: studies of polysomes, use of specific inhibitors, and cell-free protein synthesis.

b. Polysome profiles. Protein synthesis in the cytosol occurs on structures called polysomes, which are a group of polyribosomes held together by binding to a molecule of messenger RNA. Ultrastructural observations of the cytoplasm of heated cells indicated that polysomes were destroyed (Heine *et al.* 1971, Welch and Suhan 1985). Sucrose velocity sedimentation of ribosomal supernatants of heated HeLa and mouse L cells confirmed the destruction of polysomal profiles in heated cells (McCormick and Penman 1969; Schochetman and Perry 1972). The recovery of protein synthesis at 37 °C after hyperthermia was associated with the return of polysomal profiles to those found in normal cells. Partial recovery of protein synthesis in HeLa cells incubated at 42 °C was also accompanied by partial restoration of the control polysomal profile (McCormick and Penman 1969). These results indicate that the inhibition of protein synthesis after hyperthermia is intimately associated with the alteration of the polysomal profile.

d. Modification by metabolic inhibitors. To understand the mechanism of inhibition of protein synthesis by heat, experiments examining the effect of inhibitors of transcription (actinomycin D) and/or translation (cycloheximide) on the process were performed.

(i) The effect of actinomycin D. Murine leukemia L5178Y cells exposed to actinomycin D either immediately after or during a 42 °C treatment, lose the ability to recover control rates of protein synthesis upon return to 37 °C (Fuhr and Overton 1974). Similar observations were made in cells treated at 44 °C. These results are in contrast to the results with HeLa cells which could recover from the inhibition of protein synthesis induced with a treatment at 42 °C for 10 minutes, when placed at 37 °C even in the presence of

actinomycin D (McCormick and Penman 1969). However, the partial recovery of protein synthesis observed during continuous exposure to 42 °C was sensitive to the presence of actinomycin D, but could occur if protein synthesis was inhibited by cycloheximide (McCormick and Penman 1969). It was concluded from these experiments that a species of RNA must be synthesized for cells to achieve partial restoration of protein synthesis at 42 °C, but that this RNA does not appear to produce a protein. This RNA was hypothesized to act by promoting an increase in the association of ribosomes to mRNA, possibly by acting as part of the initiation process.

(ii) The effect of cycloheximide. Cells incubated in the presence of low levels of cycloheximide after an initial 10 minutes at 42 °C (which collapsed the majority of the polysomes) have polysome profiles similar to those of unheated cells (McCormick and Penman 1969). The low doses of cycloheximide slow the rate of translation, leading to an increase in the loading of ribosomes. These results indicate that the capacity of ribosomes to be engaged in protein synthesis was not affected by the heat treatment. Other experiments indicated that the size of polypeptides synthesized at 42 °C was not significantly smaller than those synthesized at 37 °C. Thus hyperthermia induced neither an aberrant functioning nor a premature release of ribosomes from mRNA. Taken altogether, these data suggested that the most probable explanation for the inhibition of protein synthesis at 42 °C was a decreased rate of initiation of protein synthesis.

Preincubation of HeLa cells with high levels of cycloheximide at 37 °C prior to exposure to hyperthermia, prevented the 42 °C-induced destruction of polysomes and inhibition of protein synthesis (McCormick and Penman 1969). The dose dependency of this phenomenon showed that the cycloheximide induced resistance to elevated temperature depended on the extent of protein synthesis inhibition during the incubation. This effect, however, as in the case with recovery of protein synthesis at 42 °C, was sensitive to actinomycin D. Thus, both the recovery process at 42 °C and the cycloheximide-induced resistance at 37 °C appear to require the synthesis of a new species of RNA, which was proposed to be a regulator of protein synthesis (see below). Pretreatment of mouse leukemia L5178 cells at 37 °C with cycloheximide prior to hyperthermia protected them from the hyperthermia-induced inhibition of protein synthesis (Fuhr *et al.* 1974). However, in this case, the presence of the inhibitor was required during the exposure to hyperthermia. Similar results were obtained with puromycin and NaF. This protection effect was sensitive to actinomycin D, as noted for HeLa cells.

e. Cell free extracts and initiation factors. In order to further examine some of the conclusions concerning the heat-induced inhibition of protein synthesis, experiments have been also performed in cell-free protein synthesizing systems.

(i) 'Initiation RNA'. A series of studies was undertaken in an attempt to define the RNA species which was postulated to play a role in promoting

the initiation of protein synthesis (Goldstein and Penman 1973; Reichman and Penman 1973; Goldstein *et al*. 1974). 1979. Crude cytoplasmic extracts prepared for HeLa cells actively incorporated amino acids but showed little initiation of new peptides, as assayed by the incorporation of N-terminal amino acids. In contrast, extracts prepared from cells subjected to prior inhibition of protein synthesis by cycloheximide showed a significant amount of polypeptide initiation, as indicated by formation of peptides with radioactive N-terminal methionine. This response was found to be sensitive to actinomycin D, indicating the need for a new species of RNA made upon treatment with cycloheximide (Reichman and Penman 1973). The actinomycin D dose–response of inhibition of recovery of protein synthesis at 42 °C, the protective effect of cycloheximide at 37 °C to subsequent exposure at 42 °C, and the inhibition of normal protein synthesis at 37 °C by actinomycin D, were all found to be similar. This result suggested that these different phenomena are related by sharing a common regulatory molecule, presumably a newly synthesized species of RNA (Goldstein and Penman 1973). Later experiments demonstrated that the situation was more complex. Extracts from cycloheximide-treated cells had a soluble stimulating factor which was actinomycin D insensitive and a ribosome associated factor that was actinomycin D sensitive. Moreover, ribosomes in crude cytoplasmic extracts from actinomycin D treated cells lost the ability to respond to stimulated supernatants (Goldstein *et al*. 1974). These effects still remain to be explained. Further, an 'initiator RNA' has yet to be isolated.

(ii) Protein initiation factors. Our understanding of the process of protein synthesis has advanced rapidly in recent years (Ochoa 1983). The initiation of protein synthesis involves several distinct factors, called initiation factors (eIF-1–4). The inhibition of protein synthesis by hyperthermia was examined in cell-free extracts of CHO cells, which actively and accurately translated exogenous natural mRNAs and synthetic polynucleotide templates (Hutchinson and Moldave 1981). Preincubation of the extract at 42 °C for 20 minutes led to a loss of 80 per cent of the globin mRNA-translating activity. Several reactions involved in the initiation of protein synthesis were examined with the following results. Both the binding of eIF-2-Met-tRNA$_f$-GTP complex to 40 S ribosomal subunits, and the subsequent reactions with mRNA and 60 S subunits to form the 80 S initiation complex were not affected by incubation at 42 °C. The formation of a ternary complex from Met-tRNA$_f$, eIF-2, and GTP and the aminoacylation of tRNA with leucine and phenylalanine were slightly inhibited at 42 °C. The translation of exogenous natural mRNA and the aminoacylation of tRNA with methionine were markedly (70–80 per cent) inhibited by preincubation at 42 °C. The authors concluded from these observations that the inactivation of Met-tRNA$_f$ synthetase was responsible for the temperature-dependent loss of ability to translate exogenous mRNAs, due to the failure of Met-tRNA$_f$ production for the process of chain initiation.

Exposure of Ehrlich ascites tumor cells grown in suspension for 20 minutes at 43 °C led to an 80 per cent inhibition of protein synthesis (Panniers and Henshaw 1984). An 80 per cent reduction of the levels of 40 S initiation complexes was found in lysates prepared from heated cells. However, the levels of Met-tRNA$_f$ synthetase activities were similar in control and heated cells. Thus, the heat-induced inhibition of protein synthesis, in this case, was thought to involve dysfunction of eIF-2. Further work by the same authors, however, reveals a different picture (Panniers *et al.* 1985). Cell-free lysates were prepared from control cells and cells exposed to 44 °C for 20 minutes. Protein synthesis in the heat-shocked lysate was significantly inhibited. However, inhibition could be reversed by the addition of either a factor purified from an Ehrlich cell ribosomal salt wash or by highly purified rabbit reticulocyte eIF-4F. The two factors had indistinguishable chromatographic properties and were proposed to be identical. eIF-4F is the cap binding protein II and may be an mRNA discriminating factor (Ray *et al.* 1983). Moreover, reduced levels of 40 A initiation complex in heated cells were restored to normal levels by the addition of eIF-4F (Panniers *et al.* 1985). There was no explanation of how eIF-4F can affect an eIF-2 function.

The relationship of alterations in initiation factors and heat-induced inhibition of protein synthesis was examined recently in HeLa cells (Duncan and Hershey 1984). Treatment of monolayer cells for 20 minutes at 45 °C resulted in a 95 per cent inhibition of protein synthesis, while incubation between 40 ° and 43.5 °C did not cause any inhibition. This observation was in contrast to results obtained with suspension culture cells, where 42 °C led to a virtually complete shutoff (described above). Since cell-free systems from HeLa cells cannot initiate protein synthesis, the authors used a more fractionated *in vitro* protein synthetic system to determine initiation factor activities. Assays for initiation factor activities in this system revealed that heat shock inhibited the activity of eIF-2, eIF-3, eIF-4 and eIF-4F. Immunoblot analysis using monoclonal antibodies indicated that heat shock induced the phosphorylation of eIF-2α and eIF-2β and diminished the extent of phosphorylation of eIF-4B. These changes were reversed in cells which had recovered at 37 °C for 2 hours after exposure to 45 °C, a time at which total protein synthesis had recovered to almost normal levels. Mixing experiments of heat-shocked HeLa lysates with reticulocyte lysates indicated that the heat-shocked lysate contained initiation factor-modifying enzymatic activities which were dominant. Phosphorylation of eIF-2 had been observed to occur in reticulocyte lysates heated at 42 °C (Ochoa 1983). It is interesting to note that eIF-2 phosphorylation and eIF-4 dephosphorylation also were observed in HeLa cells when protein synthesis was inhibited by serum factor depletion (Duncan and Hershey 1984).

More recently, the phosphorylation of eIF-2a was reported in suspension grown HeLa cells exposed to 42.5 °C which inhibited protein synthesis (DeBenedetti and Baglioni 1986). In cell-free extracts, protein synthesis inhibition was reversed if hemin was added to the lysates. In rabbit reticulocytes hemin is known to inhibit a protein kinase which represses protein synthesis by phosphorylation of eIF-2 (Ochoa 1983). By analogy, hyperthermia appears to activate a protein kinase (sensitive to hemin), responsible for inhibition of protein synthesis via phosphorylation of eIF-2.

f. The cytoskeleton and protein synthesis. Several lines of evidence have indicated that the cytoskeleton may play a structural role in protein synthesis (Lenk *et al.* 1977; Cervera *et al.* 1981; Van Venrooij *et al.* 1981). Immunoblotting experiments revealed that eIF-2, 3, and 4 were all enriched in the cytoskeletal fraction of HeLa cells (Howe and Hershey 1984). After heat shock of 7 minutes at 43 °C, the initiation factors were released from the cytoskeleton. It would be of interest to determine whether this dissociation is related to the alterations in phosphorylation of eIF's in heated cells.

4. Summary

Hyperthermia induces alterations in both the structure and function of cytosol elements. Morphological changes are observed in the mitochondria which may be related to the inhibition of respiration and glycolysis in heated cells. The membranous organelles, such as the Golgi and endoplasmic reticulum are also affected.The functional consequences of these structural changes remain to be established. Polysomes are destroyed and protein synthesis is inhibited. The inhibition of protein synthesis occurs at the initiation step and may be mediated via the phosphorylation of initiation factors. The association of polysomes and initiation factors with cytoskeletal structures is also altered, indicating the attractive possibility of a functional correlate between the heat-induced cytoskeletal changes and protein synthesis.

E. NUCLEUS

The nucleus contains the major portion of the cell's genetic information encoded in DNA. Nuclear functions involve the maintenance, duplication and expression of this information. All of these functions are disrupted or altered by hyperthermia. In addition, a number of structural changes in the nucleus have been reported (Warters and Roti Roti 1982). We will begin with the structural changes so they will be in mind when the functional changes are reviewed.

1. Structural alterations

a. Morphological alterations. Light microscopic, transmission electron microscopic and scanning electron microscopic observations have demonstrated a variety of morphological changes in the nucleus and in the nucleolus.

(i) Overall nuclear effects. One prominent feature is the condensation of cytoplasmic material into the perinuclear region (see Cytoskeleton) accompanied by an increased vessiculation of the nuclear membrane (Heine *et al.* 1971; Welch and Suhan 1985; Warters *et al.* 1986). Nuclei in heated HeLa cells show decreased heterochromatin content (Warters *et al.* 1986) and an increased number of perichromatin granules (Heine *et al.* 1971). In rat

fibroblasts actin bundles are observed within the nucleus after heating (Welch and Suhan 1985). However, the diameter of the HeLa nucleus remains unaltered after heat exposures up to 45 °C for 30 minutes (Blair *et al.* 1979; Warters *et al.* 1986). Isolation of the nucleus results in a loss of vessiculation in nuclei from both control and heated cells; but the decondensation of the heterochromatin is still evident (Warters *et al.* 1986). In addition, the nuclear peripheral lamina region appears to be retained more completely in nuclei isolated from heated cells. This latter observation may reflect the presence of heat-induced excess nuclear proteins.

(ii) Nucleolar effects. One of the distinct substructures within the nucleus is the nucleolus. This structure is the site of transcription and processing of ribosomal RNA (rRNA) (Alberts *et al.* 1983). Electron microscopic studies reveal that unlike cytoplasmic organelles, the nucleolus has no membrane. The nucleolus seems to be constructed by the specific binding of unfinished ribosome precursors to each other by means as yet unknown. Three partially segregated regions can be distinguished: (1), a pale staining component, which contains DNA from the nucleolar organizer region of the chromosomes; (2), a granular component, which contains 15 nm-diameter particles representing the most mature ribosomal precursor particles i.e. ribonucleoprotein (RNP) particles, and (3), a dense fibrilar component composed of many fine, 5 nm ribonucleoprotein fibers, representing rRNA transcripts.

Incubation of BHK cells at elevated temperatures (42–45 °C) caused the disappearance of the nucleolar RNP granules and internucleolar chromatin (Simard and Bernhard 1967). The first observable changes occurred after 1 hour at 41 °C in a few nucleoli. After 15 minutes at 42 °C the fibrilar reticulum totally disappeared and the granular form of RNP was completely lost. Similar lesions in the nucleolus were observed if the cells were incubated at 43 °, 44 ° or 45 °C. In cells returned to 37 °C for 24 hours after 1 hour at 42 °, 43 ° or 44 °C, the nucleolus resumes its normal appearance, except that the granular RNP exists in a larger than normal amount, indicating increased metabolic activity. Similar changes were observed in normal diploid rat embryonic cells exposed to 42 °C (Simard and Bernhard 1967). Nucleoli of hepatoma ascites tumor cells incubated at 44.5 °C for 30 minutes display the disappearance of the granular form of their RNP particles associated with a retraction of intranucleolar chromatin (Amalric *et al.* 1969; Simard *et al.* 1969). Kinetic studies indicate that the retraction of intranucleolar chromatin precedes the configuration changes of the granular RNP's and is associated with a rapid decrease of the specific activity of nucleolar RNA. High resolution autoradiography indicates the absence of incorporation of tritiated uridine in heated cells. If the incorporation precedes the thermic shock, radioactivity accumulates in the nucleolus and does not migrate to the cytoplasm, indicating a possible sequestering of nucleolar RNA (see below). Thus, the nucleolus is a very heat sensitive organelle, undergoing marked changes at heat exposures which leave cytoplasmic organelles largely unaffected (Simard and Bernhard 1967).

b. Alterations reflected in biochemical assays. Structural alterations in nuclei can be observed using a variety of biochemical techniques on isolated nuclei, subnuclear particles and/or model systems. These include increased nuclear protein content (Tomasovic *et al.* 1978; Roti Roti and Winward 1978) which is associated in part with the nuclear matrix (Warters *et al.* 1986); reduced access of nuclear DNA to the enzymatic probes Microccocal (M.) nuclease (Warters *et al.* 1980) and DNase I (Roti Roti *et al.* 1985); reductions in the activity of DNA and RNA polymerases (Spiro *et al.* 1982, Caizergues-Ferrer *et al.* 1980); sequestering of some nuclear matrix-associated RNA from RNase digestion (Wright *et al.* In press); and denaturation of nucleosome DNA (Seligy and Poon 1978). This latter effect requires temperatures higher than the 42 °C–46 °C range in which most of the hyperthermic effects discussed in the present review are observed.

(i) Excess nuclear protein. The increased nuclear protein content observed following hyperthermia is a large and rapid effect. The effect correlates with heat-induced cell killing (Roti Roti *et al.* 1979; also see Figs 4 and 5). In fact, it is the earliest effect after heat shock which correlates with cell killing.

Detection. The presence of excess nuclear protein (Fig. 1) was first observed in isolated chromatin (Tomasovic *et al.* 1978; Roti Roti and Winward 1978) using chemical measurements of protein and DNA or measuring the radioactivity of prelabeled protein (via ^{3}H-amino acids) and DNA (via ^{14}C-TdR). The chromatin preparations used in these studies contained both the nuclear matrix and the nucleosomes. Some later studies (e.g., Wheeler and Warters 1982) refer to the nucleosomes as chromatin distinguishing it from the nuclear matrix. Lack of awareness of this distinction can cause some confusion in the interpretation of results. Subsequent studies showed that the excess nuclear protein could be measured by flow cytometry (FCM) using isolated nuclei stained with the fluorescent dye, fluorescein isothiocyanate (FITC), (Blair *et al.* 1979). This method was expanded to include simultaneous measurement of nuclear DNA content (Roti Roti *et al.* 1982). The FCM methods have two advantages over previous methods: (1), the ability to obtain data rapidly, and (2), the ability to obtain data from individual nuclei. This latter characteristic was exploited to show that when nuclei are isolated from a mixture of unheated and heated cells the protein contents of nuclei of unheated and heated cells remain unaltered (Roti Roti *et al.* 1984).

Localization. Several studies suggest that a significant portion of the heat-induced excess nuclear protein is associated with the nuclear matrix. The nuclear matrix is operationally defined as the residual particle remaining when the histones and DNA are removed. It consists of the nuclear lamina and an internal fibrilar network and is thought to be the site of DNA replication, RNA transcription, and possibly DNA repair (Maul 1982). Nucleoids consist of dehistonized DNA loops and the nuclear matrix. These particles have an increased protein content following hyperthermia (Roti

Roti and Painter 1982). Limit digestion of nuclei from heated cells with DNase I or M. nuclease solubilized no detectable amount of the excess nuclear protein; instead the excess nuclear proteins co-sedimented with the nuclear matrix (Wheeler and Warters 1982). Direct observation of nuclear matrices by filter-trapping assays or sedimentation assays (Warters *et al.* 1986) and FCM assays (Wright *et al.*) showed an increased amount of protein associated with matrices from heated cells. The measured amount of the heat-induced protein appears to be dependent upon the order in which nuclei are fractionated (Table 1). The measured protein content of the matrix from the heated cells was least when the 2M NaCl soluble proteins were removed first. This observation suggests that these excess proteins interact with the nuclear matrix via other macromolecules. Since the extraction of nucleic acids is the variable in the preparation, presumably these are the macromolecules involved in the interaction. This notion is supported by the

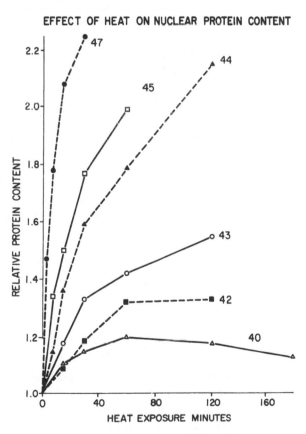

Figure 1: Effects of heat on nuclear protein content. HeLa cells were exposed to the various time–temperature combinations indicated on the graph, their nuclei isolated and stained for DNA and protein content according to published methods (Roti Roti *et al.* 1982). The relative (control = 1) nuclear protein content is plotted versus heating time for each temperature. The plotted points represent the mean of at least three repeated experiments; error bars have not been omitted for clarity.

maximization of apparent sequestering of RNA by these excess nuclear proteins when the 2M NaCl extraction is the last step (see Fig. 3).

Characterization. Although significant progress has been made in determining the localization of the heat-induced nuclear protein, relatively little is known about these proteins individually. The proteins which constitute the heat-induced excess nuclear proteins have proved refractory to the standard solubilization techniques. This problem has caused gel electrophoretic analysis of these proteins to be difficult. Early studies showed that the excess proteins did not contain histones (Tomasovic *et al.* 1978). This result was supported by the observation that the excess nuclear proteins were not enriched in lysine (Higashikubo *et al.* 1986), as were the histones. In contrast the excess proteins appeared to be enriched for leucine and tryptophan, and a fraction of these were turning over rapidly (Higashikubo *et al.* 1986). Immunofluorescence studies have shown that certain proteins are translocated into the nucleus by heat shock. One of these, the 70 kD heat-shock protein (HSP), is associated with the nucleolar region (Welch and Feramisco 1984; Pelham 1984). Two-dimensional gel electrophoresis of proteins from nuclei of heat-shocked cells which had been prelabeled with ^{35}S-methionine for 24 hours showed an increased amount of HSP 70 but no change in cytoskeletal proteins (Laszlo *et al.* unpublished). Although HSP may contribute to the excess nuclear protein content, this response is clearly distinguishable from the induction of HSP synthesis. Inhibition of RNA or protein synthesis by pretreating cells with Act-D or CHM (respectively), for 1 hour prior to heat-shock did not significantly alter the relative increase in nuclear protein content (Higashikubo *et al.* 1986). This observation suggests that the proteins which constitute the heat-induced excess nuclear protein are present in the cell at the time of heat shock. However, the HSP are synthesized as a result of heat shock (Schlesinger *et al.* 1982). The recently reported 'prompt heat-shock proteins' could be a part of the heat-induced excess nuclear proteins (Reiter and Penman 1983). However, the 'prompt-heat-shock' protein can only be a small fraction of the total, because the mass change due to them was at the limits of detection.

(ii) Enzymactic access to nuclear DNA. Certain changes in chromatin structure can be detected by the sensitivity of nuclear DNA to the enzymatic probes, M. nuclease and DNase I (Noll 1974).

M. nuclease. In the eukaryotic nucleus DNA is packaged in association with histones to form particles known as nucleosomes. Nucleosomes, in effect, divide the DNA into core regions (i.e. tightly associated with histones) and linker regions (i.e. loosely associated with histones). M. nuclease preferentially digests the internucleosomal or linker DNA resulting in a population of mononucleosomes at limit digest (i.e. a population of DNA fragments of 140–160 bp in length) (Kornberg 1977; Chambon 1978). Exposure of HeLa cells to hyperthermia (43.5–48 °C for 30 minutes) causes a temperature dependent reduction in the rate of linker DNA solubilization (Warters *et al.*

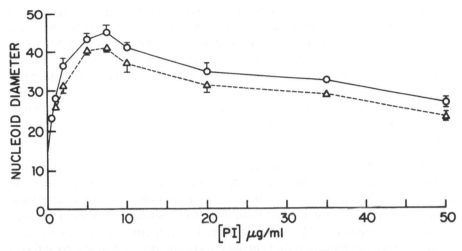

Figure 2: Effects of heat on the DNA available for supercoiling changes. DNA supercoiling changes were measured by the fluorescence halo assay (Roti Roti and Wright 1987). Nucleoids were made from HeLa cells by lysing cells in the presence of 1 M NaCl and the indicated Propidium iodide (PI) concentration (abscissa) for 30 minutes. The resulting fluorescent halo consisted of DNA loops extending beyond the nuclear lamina. Various degrees of DNA supercoiling were reflected in the overall halo diameter in microns (ordinate). The symbols (open circles, control and open triangles, 45 °C) represent the mean of three repeated experiments and the bars represent one standard error (omitted when smaller than the symbol).

1980). However, these hyperthermic exposures did not alter the fraction of total DNA solubilized at limit digestion, the modal size of the core DNA, or the distribution of oligonucleosome sizes at equivalent solubilization levels. These observations suggest that hyperthermia affects the initial stages of M. nuclease digestion. Newly replicated (nascent) DNA in chromatin can be distinguished from parental DNA in chromatin by its greater sensitivity to M. nuclease (Seale 1975; Weintraub 1976), which is reflected in both a more rapid initial digestion rate and a greater fraction of DNA solubilized at limit digestion. Hyperthermic exposure (43.5–48 °C, 30 minutes) inhibited in equal proportion the digestion rate of both nascent DNA and parental DNA in chromatin, but did not affect the fractions of DNA solubilized at the limit digest (Warters and Roti Roti 1981). The sedimentation coefficient and buoyant density of nucleosomes from both nascent and parental chromatin were unaffected by hyperthermia (45 °C, 60 minutes) (Warters and Stone 1984a). However, at limit digestion in heated cells, there was significant increase in DNA fragments of subnucleosomal size (≈ 125 bp) without altering the mode (145 bp) of the core DNA size distribution (Warters and Stone 1984a). This result suggested that heat allowed M. nuclease to make intercore cuts in a small fraction of the nucleosomes.

DNase I. DNase I has both an endonucleolytic and an exonucleolytic action. These modes of action can be distinguished in nuclei by FCM analysis of the digestion kinetics using ethidium bromide fluorescence (Darzynkiewicz *et al*. 1981; Roti Roti *et al*. 1985). The endo activity is seen as an increase in fluorescence due to nicks in the DNA which allow a greater amount of ethidium bromide binding, while the exo activity is seen as loss of fluorescence due to the solubilization of DNA. The limit of the digestion appears to reflect the fraction of DNA in close association with the nuclear matrix (Wheeler and Warters 1982). Hyperthermia (45–47 °C, 30 minutes) does not affect the initial DNA nicking, but does inhibit the rate of DNA solubilization and increases the fraction of DNA remaining at the limit digest in a temperature dependent manner (Roti Roti *et al.* 1985). Thus hyperthermia appears to affect the later stages of the DNase I digestion process.

Mechanism for enzymatic inhibition. In the above studies, the enzyme is added exogenously; therefore, the alterations in kinetics must be due to an alteration in the chromatin substrate. The inhibition of nuclease digestion cannot be due to thermal denaturation of nucleosomal DNA because such denaturation facilitates M. nuclease digestion (Dimitrov *et al.* 1980). The most probable mechanism is that changes in chromatin-nuclear matrix interactions, associated with the heat-induced increase in nuclear protein content cause these alterations in enzyme kinetics. The linear correlation between the inhibition of M. nuclease digestion and the amount of excess nuclear protein is evidence for this notion (Warters *et al.* 1980). However, digestion of nuclear DNA with either M. nuclease or DNase I does not release any of the excess nuclear proteins (Wheeler and Warters 1982), suggesting that the binding of these proteins is not dependent upon nucleosome associated DNA. This observation is consistent with the manner in which hyperthermia inhibits DNAse I digestion, but is apparently inconsistent with that for M. nuclease. This inconsistency can be resolved beginning with the fact that nuclear DNA is organized in loops which are attached to the nuclear matrix (Vogelstein *et al.* 1980). These loops, known as domains, contain DNA organized into nucleosomes. The heat-induced excess proteins associated with the nuclear matrix reduce the ability of the DNA in the domains to undergo changes in supercoiling (Roti Roti and Painter 1982; and see Fig. 2). Therefore, if one assumes that M. nuclease digestion of the linker DNA requires some rotation of the DNA throughout the loop, then the apparent inconsistency is resolved.

(iii) Inactivation of DNA and RNA polymerases: Thermodynamic analysis of cell killing led to the proposition that protein denaturation is a rate limiting step in the lethal process (Westra and Dewey 1971). Subsequent work has shown that the enzymatic activity of several critical polymerases are heat sensitive and/or depend upon heat-sensitive components.

DNA polymerases. Mammalian cells contain three nuclear DNA polymerases designated as polymerase-α -β, and -γ (Loeb 1974; Bollum 1975; Weissbach 1975). Polymerase-α is believed to be the enzyme involved in DNA replication because it increases as cells initiate DNA replication (Bollum 1975; Loeb 1975; Weissbach 1975). Further, the level of polymerase-α activity is decreased as temperature sensitive mutants are held at nonpermissive temperature; its increase is tightly coupled with re-entry into S-phase upon shift to permissive temperature, while polymerase-β and polymerase-γ remain constant (Schneider *et al.* 1985). Polymerase-β is believed to function in DNA repair (Bollom 1975; Bertazzoni *et al.* 1976). The function of polymerase-γ is unclear and little data are available on the effects of heat on this enzyme. Of the other two enzymes, polymerase-β is more heat sensitive than polymerase-α. When heated and assayed *in vitro*, polymerase-α retains 70 per cent of its activity after 40 minutes at 45 °C (Dube *et al.* 1977), whereas polymerase-β is inactivated to less than 19 per cent of its original activity within 6 to 35 minutes at 45 °C, depending on the source and method of isolation of the enzyme (Spiro *et al.* 1983; Dube *et al.* 1977). When the enzymes are heated *in situ* and then analyzed in frozen-thawed cells, both polymerases show a heat-induced loss of activity (Spiro *et al.* 1982). Under these conditions, degree of the heat-induced inactivation of polymerases-α and -β is reduced in the presence of the thermoprotector, glycerol, (Mivechi and Dewey 1984, 1985) and is increased in the presence of the thermosensitizer, procaine, (Spiro *et al.* 1982). However, procaine has no effect on the thermal sensitivity of polymerase-β heated as an isolated enzyme (Spiro *et al.* 1983). These results suggest two thermal effects on mammalian DNA polymerases *in situ*, direct thermal denaturation of the enzymes and scquestering of the enzymes. In the case of polymerase-β, it appears as if both mechanisms contribute to the loss of enzyme activity, while for polymerase-α only the second mechanism appears to make a significant contribution to the loss of activity. Since these polymerases are known to be associated with the nuclear matrix, it is not inconceivable that the sequestering mechanism involves heat-induced excess nuclear proteins. When these enzymes were measured *in situ* in isolated nuclei, it was found that hyperthermic treatment of cells increased the activity of both enzymes (Kampinga *et al.* 1985) showing that they can be stabilized in the nucleus. The increase in polymerase-α activity was greater than that of polymerase-β consistent with the above considerations.

RNA polymerases. There are three RNA polymerases in mammalian cells: polymerase-I, which makes large ribosomal RNA's; polymerase-II, which transcribes genes; and polymerase-III, which makes a variety of small RNA's (Chambon 1978). After heating CHO cells at 43 °C for 60 minutes, the remaining enzyme activities were: 0.50 of polymerase-I, 0.81 of polymerase-II, and 0.76 of polymerase-III, (Caizergues-Ferrer *et al.* 1980). For comparison, a similar heat exposure will reduce the activity of DNA polymerase-β to 0.27 (Spiro *et al.* 1982). Thus, the heat sensitivity of RNA polymerases is half or less than that of DNA polymerase-β and comparable to that of DNA polymerase-α.

(iv) Thermal denaturation of nucleosomal DNA. DNA denaturation can be determined by changes in absorption of 260 nm light (Klump 1977) or by changes in the binding of absorbing (Dreskin and Mayall 1974; Roth 1978) or fluorescent (Darzynkiewicz *et al.* 1975) dyes. The former technique

is primarily used for studying denaturation of isolated DNA either free or in nucleo-protein complexes in solution, whereas the dye binding assays are used for *in situ* DNA denaturation studies. The two approaches have not been experimentally correlated.

Absorption studies. Bulk mammalian DNA isolated from cells has 3 to 10 transition temperatures in the denaturation curves, depending upon the source of the DNA, the salt concentration, and pH of the solution (Klump 1977; Lewis 1977; Defer *et al.* 1977; Guttmann *et al* 1977). All of these transitions occurred at temperatures above 60 °C. If the DNA is in mononucleosome-sized fragments, then the denaturation curves have about three transition temperatures beginning at about 40 °C in 6M urea (Seligy and Poon 1978). However, when the DNA is associated with histones in a mononucleosome, the minimum transition temperature is increased to 50 °C and higher. This minimum transition temperature does not appear to vary if the size of the chromatin fragment is increased to dimer or trimer size or if the nucleosomes are acetylated (Yau *et al.* 1982). The presence of HMG-1, a nonhistone chromatin protein, increased the transition temperatures of the DNA melting curves (Butler *et al.* 1985). In sheared whole chromatin (i.e. containing both histone and nonhistone protein) or chromatin fractions (e.g., trypsinized, dialyzed, etc.) all of the DNA transition temperatures were higher than 50 °C (Defer *et al.* 1977; Lewis 1977). Thus, the model system studies suggest that DNA in most possible *in situ* conformations does not become denatured at temperatures encountered in hyperthermia studies. However, if whole chromatin is studied, a small fraction of the DNA has a transition in the 40 °C to 50 °C range (Lewis 1977; Dimitrov *et al.* 1980). Clearly this small fraction of heat sensitive DNA has yet to be duplicated in any of the model systems. More importantly, the question remains whether or not detectable DNA denaturation occurs *in situ* in the 40 °C to 48 °C range.

Dye binding studies. The approach used in these studies allows the investigation of DNA conformational changes *in situ*. In general, the assays determine the fraction of denatured DNA via the different binding modes or affinities of the dyes for single and double stranded DNA. The staining of DNA by gallocyanin-chrome alum (GCA) was found to vary with the compactness of chromatin (Dreskin and Mayall 1974). Thermal denaturation (confirmed by S_1 nuclease digestion) of chromatin increased GCA binding. However, this increase in GCA binding was insignificant below 70 °C (Dreskin and Mayall 1974). When the dye acriflavin is bound to DNA, its fluorescence is quenched at 4 °C. The increase in fluorescence of acriflavin as it dissociates from unwinding DNA can be used as an assay for DNA denaturation (Roth 1978). DNA denaturation detectable by this method was insignificant at temperatures below 58 °C (Roth 1978). Another approach is to use the bimodal binding properties of the fluorescent dye, Acridine Orange (AO). AO intercalated into double stranded DNA emits green fluorescent light; whereas AO bound to single stranded nucleic acid emits red light. Thus, the relative portion of total fluorescence that is red can be used to indicate the state of denaturation of DNA (Darzynkiewicz *et al*. 1975). This approach, allowed demonstration of a low temperature (40 °C to 50 °C) transition of DNA denaturation *in situ* involving 10 per cent or less of the total DNA (Darzynkiewicz *et al*. 1975) depending on the cell type, and that the fraction of low temperature denaturing DNA was increased by 50 per cent in mitotic cells (Darzynkiewicz *et al*. 1977). Interestingly, the removal of HMG proteins also increased the fraction of DNA undergoing the low temperature transition (Darzynkiewicz *et al*. 1976).

Effects in the hyperthermic range 40 °–48 °C. Most of the results of the above studies are obtained at temperatures above the 40 °–48 °C range. However, a few studies demonstrate the presence of a small fraction of DNA undergoing low-temperature denaturation. In native chromatin denaturation, transitions appear at temperatures below those for free DNA, i.e. in the 40 °C to 50 °C range (Lewis 1977; Defer *et al*. 1977). The low-temperature transitions involve a small fraction ($\approx$ 10 per cent) of the total DNA. These transitions are reduced or eliminated if the chromatin is dialyzed or sheared (Lewis 1977), or if all the histones are removed (Defer *et al*. 1977). However, if histone H1 is removed, the fraction of DNA undergoing low-temperature denaturation is increased (Dimitrov *et al*. 1980), suggesting that certain configurations of chromatin hold regions of the DNA in a thermally sensitive state. The low-temperature DNA denaturation transitions are observed also *in situ* (Darzynkiewicz *et al*. 1975). This result, along with the observation that M. nuclease nicks core DNA into sub-mononucleosome size lengths in a small fraction of nucleosomes from heated cells (Warters and Stone 1984a), suggests that hyperthermia causes some loosening of normal DNA–nucleoprotein structure in a small fraction of the chromatin.

(v) DNA degradation. Following heat doses of 45 °C for up to 180 minutes no DNA degradation (i.e. measured as TCA solubilization of DNA) can be observed immediately following hyperthermia (Warters *et al*. 1985). Temperatures of 48 °C or greater were required to observe DNA degradation. However, if HeLa cells were exposed to 45 °C for 30 minutes and observed 24 hours later, a significant DNA degradation is observed (Warters 1982). However, in monolayer culture, the late developing DNA degradation is associated with interphase death rather than the early stages of reproductive death (Warters and Henle 1982).

(vi) RNA associated with the nuclear matrix. The work reviewed below is unpublished (except for the FCM Methodology, Wright *et al*. in press) and should be considered work in progress. We are including these results for completeness and because they appear to be a novel and important nuclear effect of hyperthermia. Nuclear matrices are made from nuclei by DNase I and RNase digestion and 2 M NaCl (HSB) washing in various sequences (Maul, 1982). The isolated nuclear matrix contains 16 to 18 per cent of the nuclear double-stranded nucleic acid (PI binding, Table 1), but only 0.1–0.01 per cent of the nuclear DNA (detected by bulk labeling). Therefore, the residual PI fluorescence, which is sensitive to RNase digestion, is due to

double stranded RNA. When such matrices are obtained from HeLa cells
heated at 45 °C for 30 minutes, the double stranded RNA measured as PI
fluorescence is less sensitive to digestion by RNase A (Fig. 3). The amount
of differential sensitivity to RNase A digestion depends upon the method of
matrix isolation (Table 1 and Fig. 3), suggesting that some structural rearrangements could be occurring during nuclear matrix fractionation.

2. Functional alterations in nuclei

a. Effects on DNA replication. In mammalian cells, DNA replication can
be conceived as occurring in four steps: (1), initiation of a cluster of replicons;
(2), elongation of the synthesized DNA chains; (3), assembly of the newly
replicated DNA with histones into nucleosomes; and (4), ligation of the
complete replicons into cluster size DNA lengths and ligation of cluster size

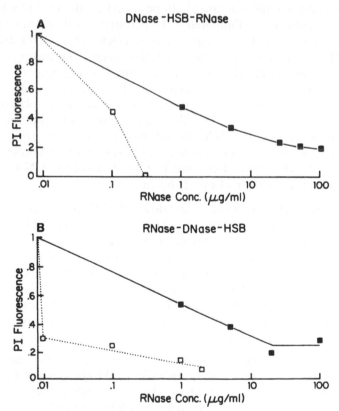

Figure 3: RNase tolerance curves for matrix associated RNA. The accessibility of double
stranded RNA (i.e. PI binding) to digestion by RNase A is detected by the assay illustrated
here. The relative PI fluorescence (intact matrix = 1) is plotted as a function of the RNase
concentration used for a 30-minute digestion at the indicated step in matrix preparation. The
solid symbols represent the matrices from heated (45 °C for 30 minutes) HeLa cells, while the
open symbols represent matrices from control cells. When the symbol is plotted at zero, the
matrices are dissociated as a result of the RNase digestion. The data are average values from
three or more experiments (from Wright *et al.* in press).

Table 1.

A. Fraction of intact nucleus*

	DNase I treated		Nuclear matrix	
	Control	45 °C–30'	Control	45 °C–30'
Double stranded (PI) nucleic acid	0.26	0.31	0.18	0.16
Protein (FITC)	0.36	0.72	0.08	0.40
Light scatter	0.43	0.43	0.25	0.25

B. Method of preparation

Extraction sequence	Fraction nuclear protein	Relative change due to heat shock	Sensitivity† to RNase	Protection by heat shock
RNase A → DNase I → HSB	0.09	7.3	+++	+++
DNase I → HSB → RNase A	0.08	6.6	+++	+++
DNase I → RNase A → HSB	0.08	6.6	+	++
HSB → DNase I → RNase A	0.07	3.0	*	*

* Results from a typical experiment showing increased retention of protein with the nuclear matrix due to heat shock. Note that matrices from both heat-shocked and control cells retain the same fraction of double stranded nucleic acid
† As illustrated in Figure 3
Results are from typical experiments
The method of isolation appears to affect the sensitivity of the nuclear matrix to RNase A but in all cases the nuclear matrices from heat-shocked cells were more resistant to RNase A and had a high protein content. (* Not determined to date)

lengths into 'chromosome size' molecules. Hyperthermia inhibits all of these processes to varying extents. The incorporation of ^{3}H-thymidine (TdR) into DNA post-hyperthermia can reflect, in part, the inhibition of all of these steps (Wong and Dewey 1982; Warters and Stone 1983a). The inhibition of ^{3}H-TdR incorporation immediately following hyperthermia showed two phases of inactivation as a function of heat dose (Warters and Stone 1984b). The rapidly inactivated component was believed to be replicon initiation, while the slowly inactivated component was believed to be elongation. We will consider the effects of heat on each of these in order.

(i) Initiation. DNA replication occurs in distinct units called replicons. A group of replicons, known as a cluster, initiate replication as a unit dependent upon DNA supercoiling changes. Analysis of nascent DNA fragments made at various time intervals after hyperthermia (45 °C, 15 minutes) by velocity gradient sedimentation has demonstrated the depression of elongation and initiation. With post-hyperthermia incubation (longer than 2 hours) elongation began to resume, but initiation was still inhibited (Wong and Dewey 1982). In fact, repression of initiation continues for 15 to 18 hours in the CHO cells (Wong and Dewey 1982) and for more than 8 hours in HeLa cells heated at 45 °C for 30 minutes (Warters and Stone 1983b). Thus, repression of

replicon initiation is a relatively long lasting heat effect. Further, replicon initiation in HeLa cells is threefold more heat sensitive ($D_0 = 7$ minutes at 45 °C) than elongation ($D_0 = 20$ minutes at 45 °C) (Warters and Stone 1984b).

(ii) Elongation. DNA chain elongation is more heat resistant than initiation and it appears to recover more rapidly. A method to measure the effects of heat on elongation is to pulse label cells with ^{3}H-TdR prior to the heat exposure and monitor the size distribution of labeled nascent DNA at various time intervals after hyperthermia (Wong and Dewey 1982; Warters and Stone 1983b). The size distribution can be measured by alkaline sucrose gradients (Wong and Dewey 1982; Warters and Stone 1983b) or alkaline elution (Warters and Stone 1983a). At 37 °C, the shift in the size of nascent DNA progressed from small fragments ($\simeq$ 20 S) through replicon size ($\simeq$ 40–60 S) to cluster size ($\simeq$ 120 S) in about 60 minutes. If cells were exposed to heat, 45.5 °C for 15 minutes for CHO (Wong and Dewey 1982), or 45 °C for 15 minutes in HeLa cells (Warters and Stone 1984b), this time course increased to 120 to 135 minutes. However, elongation is not completely stopped until heat exposures of 45 °C for 60 minutes (Warters and Stone 1983b). Following a dose of 45 °C for 15 minutes in HeLa cells, 6 hours were required before elongation returned to its normal rate (Warters and Stone 1984b). Elongation recovery time appears to be somewhat longer in CHO cells exposed to 45.5 °C for 15 minutes (Wong and Dewey 1982).

(iii) Assembly of nascent DNA into chromatin. Although the DNA double helix is bidirectional, DNA replication is unidirectional. This fact leads to the situation that replication is continuous on the leading (5' phosphate) strand and discontinuous on the trailing (3' hydroxyl) strand (Alberts *et al.* 1983). The parental nucleosomes are thought to remain loosely associated with the leading strand while the newly made nucleosomes become associated with the trailing strand (Weintraub 1976). Thus, at the growing point, the DNA is dissociated in part from the nucleosomes, which makes the most recently replicated DNA sensitive to digestion by M. nuclease (Seale 1975). Relative to parental DNA, a higher fraction of nascent DNA is digested by M. nuclease and the digestion kinetics are more rapid. If assembly of nascent DNA into chromatin is followed via a pulse-chase experiment, then this differential in M. nuclease sensitivity is lost with a $T_{1/2}$ of 2.5 minutes in HeLa cells (Warters and Roti Roti 1981). When HeLa cells had been exposed to hyperthermia (43.5 °C to 48 °C for 30 minutes) the $T_{1/2}$ for this conversion was increased to 21–218 minutes linearly with heat dose. Following 45 °C for 30 minutes, the $T_{1/2}$ was 81 minutes with the conversion being 85 per cent complete at 5-hour post-hyperthermia (Warters and Roti Roti 1981). Thus, hyperthermia causes a severe inhibition of chromatin assembly, resulting in a fraction of DNA remaining in a potentially unstable state for relatively long time periods.

(iv) Ligation into cluster and 'chromosome size' fragments. Ligation of replicon size pieces into cluster size pieces appears to occur as elongation is

complete. Continued ligation will produce 'chromosome size' pieces ($\simeq 200$ S). At 5 to 15 hours post-hyperthermia (45.5 °C for 15 minutes in CHO cells), most of the nascent DNA fragments are 120 S size pieces, indicating a failure to ligate into 'chromosome size' fragments (Wong and Dewey 1982). Between 15 and 18 hours post-hyperthermia, when the rates of both initiation and elongation have returned to control values, the 120 S size pieces are observed to be ligated into 'chromosome size' fragments ($\simeq 200$ S) (Wong and Dewey 1982). This result suggests that once clusters are complete, they require the initiation and completion of neighboring clusters prior to the formation of chromosome-size fragments.

(v) Mechanism of the inhibition of DNA synthesis. The above effects on DNA replication are too large to be caused by the relatively small amount of direct DNA damage (see Section E1, *b, iv* and *v*). Further, the activation energy for DNA damage induction is much lower than that for DNA synthesis inhibition (Warters and Stone 1983a). Hyperthermia in the dose ranges used in these studies does not affect the pool size of DNA precursors (Warters and Stone 1984b), nor the structure of the nucleosome (Warters and Stone 1984a). Thus, the viable possibilities are: inhibition of DNA polymerase; unavailability of histones due to disruption of protein synthesis; and nuclear protein content changes. Recent work (Kampinga *et al.* 1985), suggests that polymerase-α, which is responsible for DNA replication, is not functionally impaired *in situ* after 45 °C 30-minutes treatment. Therefore, disruption of polymerase-α can only play a small role in the inhibition of DNA synthesis. Histone synthesis returns to control levels prior to the resumption of DNA synthesis (Warters and Stone 1983b), suggesting that inhibition of histone synthesis is not responsible for the overall inhibition of DNA synthesis. However, the resumption of histone synthesis does appear to correlate with the resumption of DNA chain elongation (Warters and Stone 1984b) under the conditions studied. Since the heat-induced excess nuclear proteins are known to be associated with the nuclear matrix (Warters *et al.* 1986; Wright *et al.* in press) which is the site of DNA replication (Pardoll *et al.* 1980), it seems reasonable to speculate that the excess nuclear proteins at the nuclear matrix are involved in the inhibition of DNA replication. Further, it is known that replicon initiation requires DNA supercoiling changes (Mattern and Painter 1979) and that the presence of excess nuclear proteins is associated with restriction of DNA supercoiling (Roti Roti and Painter 1982). The removal of excess nuclear protein (Roti Roti *et al.* 1986) appears to be correlated inversely with resumption of DNA chain elongation (Warters and Stone 1983b; 1984b), while the resumption of replicon intitiation appears to occur at the time when most of the excess nuclear proteins are removed from the nucleus (Roti Roti *et al.* 1986). In heated cells the $T_{1/2}$ for chromatin assembly is correlated with the nuclear protein content (Warters and Roti Roti 1981). Thus, heat-induced excess nuclear proteins can be associated with inhibition of at least three of the 4 steps in DNA replication. However, the appropriate cause and effect experiments remain to be done before a definitive statement can be made.

c. Effects on RNA production. The synthesis of RNA is one of the major functions of the nucleus. The effects of hyperthermia on the rate of RNA synthesis has been examined in Ehrlich ascites cells (Strom *et al*. 1973), HeLa cells (Warocquier and Scherrer 1969) and CHO cells (Henle and Leeper 1979). In all cases RNA synthesis was inhibited by hyperthermia. However, in most cells, 40 per cent to 70 per cent of RNA synthetic activity is devoted to the synthesis of ribosomal RNA (rRNA) (Alberts *et al*. 1983). Therefore, the hyperthermia-induced inhibition of RNA synthesis can be accounted for in great part by the effect on the synthesis and processing of rRNA. Nevertheless, both the production of mRNA and the production of nRNA are disrupted by hyperthermia.

i. Effects on transcription and processing of hnRNA and mRNA. The synthesis of heterogenous nuclear RNA (hnRNA) and its processing into mRNA after hyperthermia in mammalian cells has not been completely characterized to date. It has been suggested that rRNA synthesis may be more heat labile than hnRNA synthesis (Warocquier and Scherrer 1969). However, the detailed kinetics of the synthesis, turnover and processing of hnRNA in heated mammalian cells remains to be examined before this idea can be confirmed.

The transcription of a specific gene, actin, is reversibly inhibited (100 per cent blocked) in heat-shocked Drosophila cells in culture within 5 minutes after the initial heat-shock (Findly and Pederson 1981). On the other hand, total incorporation into hnRNA remained at 25 per cent of control in heat-shocked cells (Mayrand and Pederson 1983). However, the assembly of the hnRNA into hnRNP is incomplete under these conditions, as evidenced by increased buoyant density of hnRNP particles in Cs_2SO_4 gradients. Direct analysis by RNA–protein cross-linking *in vivo* also revealed that the hnRNA made during heat shock was complexed with greatly reduced amounts of protein. Heat induced alterations in the hnRNP structure of HeLa cells exposed to 43 °C that were similar to those found in Drosophila (Mayrand and Pederson 1983). These alterations were observed to begin at temperatures from 39 °C to 41 °C. The consequences of the alterations in hnRNP assembly on the processes of hnRNA turnover and mRNA production remain to be determined.

(ii) Effects on transcription and processing of rRNA. The original rRNA transcript made in the nucleolus is very large, about 45 S. The transcript is first packaged into a large RNP particle containing many different proteins imported from the cytoplasm. As the 45 S RNA molecule is processed further, the large ribonucleoprotein particle gradually loses some of its RNA and protein, and then splits to form separate precursors of the large and small ribosomal subunits. The appearance in the cytoplasm of the smaller ribosomal subunit, the 40 S particle containing the 18 S rRNA, takes about 30 minutes. The appearance of the mature large ribosomal subunit, the 60 S particle containing the 28 S, 5.8 S and 5 S rRNAs, takes about 60 minutes to complete (Alberts *et al*. 1983).

Nucleolar and ribosomal RNA in heated hepatoma ascites cells were analyzed by density gradient centrifugation (Amalric *et al.* 1969). The incorporation of tritiated uridine into the 45 S peak decreased dramatically after a 30-minute treatment at 44.5 °C; if the labeling preceded the heat shock, then the radioactivity was retained in the 45 S peak. Biochemical studies were also performed on HeLa cells exposed to 42 °C (Warocquier and Scherrer 1969). This study indicated that under such conditions, the production of the 45 S pre-rRNA decreased dramatically and its processing ceased almost completely, so that no new rRNA appeared in cytoplasmic ribosomes. On the contrary, the synthesis of nascent messenger-like RNA was only slightly affected and the non-ribosomal RNA appeared in the cytoplasm at a normal rate. The changes in rRNA metabolism observed at 42 °C were reversible upon return to 37 °C. Studies performed on Hep-2 cells exposed to 42 °C revealed, in addition, that the synthesis and processing of pre-rRNA recovered later than protein synthesis (Cervera 1978). This observation suggested that both the synthesis and processing of rRNA transcripts depends on protein synthesis.

Further information on the effects of heat on nucleoli would be useful, especially in the light of reports that the induced form of the 70 kD HSP family is associated with the nucleolus in rat cells exposed to 42 °C (Welch and Feramisco 1984), and that the 110 kD HSP is localized in the nucleolus of non-heated mouse and hamster cells (Subjeck *et al.* 1983). The role of the association of these HSPs with the nucleolus is not clear at this time. It would be most interesting to determine whether any temporal relationship exists between the association of these HSP with the nucleolus and the functional changes occurring in the nucleoli, such as alterations in the transcription and processing of rRNA. Indeed, evidence has been presented that the 70 kD HSP may play a role in the recovery of nucleolar morphology after exposure to hyperthermia (Pelham 1984).

c. Effects on DNA repair. Hyperthermia is known to interact synergistically with ionizing radiation in terms of cell killing (Dewey *et al.* 1978). Such an interaction can be due either to a potentiation of damage or the inhibition of repair. Hyperthermia does not cause significant potentiation of cells to radiation-induced DNA damage of the following types: single strand breaks (Clark *et al.* 1981; Mills and Meyn 1981; Jorritsma and Konings 1984); double strand breaks (Corry *et al.* 1977); thymine base damage (Warters and Roti Roti 1979); and DNA–protein cross links (Cress and Bowden 1983). The major effect in cells exposed to hyperthermia is the loss of the ability to repair these types of damage. Since later chapters in this series will cover heat and X-ray interactions, we will emphasize the effects of heat on the various DNA repair processes. Although current state of knowledge is such that documentation of heat effects on the various steps for any repair process is incomplete, certain trends can be seen from the available data.

(i) Excision of thymine base damage. The production and excision of products of the 5′,6′-dihydroxydihydrothymine type (T′) were studied in

CHO cells. It was shown that hyperthermia (45 °C, 15 minutes) did not affect production of these products by X-irradiation, but did inhibit excision of the products post-irradiation (Warters and Roti Roti 1979). Further studies showed that sonicates prepared from heated (45 °C, 15 minutes) and control cells were equally capable of excising T' products from X-irradiated DNA or chromatin; whereas sonicates of control cells were inhibited in their ability to excise T' products from X-irradiated chromatin from heated cells (Warters and Roti Roti 1979). These results show that hyperthermia did not significantly denature the enzymes responsible for excising T' products. Therefore, the inhibition of excision repair was due to a heat-induced alteration of the substrate, i.e. DNA in chromatin. In this study chromatin from heated cells contained 50 per cent more protein than that from control cells, suggesting that inhibition of T' excision may be related to the excess nuclear proteins.

(ii) Rejoining of DNA strand breaks. The heat-induced inhibition of DNA strand break rejoining appears to be due to heat-induced alterations in chromatin structure. However, the evidence comes from different types of experiments. In heterokaryons made from irradiated cells and heated and irradiated cells by cell fusion, there was strand break rejoining in the former nuclei, but no rejoining in the latter nuclei (Clark and Lett 1978). This result suggested that an alteration in the nucleus of the heated cells was responsible for the inhibition of strand break rejoining. Further studies showed that both the extent of inhibition of rejoining and the extent of radiosensitization correlated with the amount of excess protein associated with chromatin, and that cells did not regain their full ability to rejoin breaks until after the excess protein had been removed (Clark *et al.* 1981; Mills and Meyn 1981). The residual DNA damage remaining after 8 hours post-irradiation repair following combined heat (43 °C for 60 minutes) plus X-ray exposures with various time intervals between the modalities was determined in CHO cells (Mills and Meyn 1983). The residual damage was reduced with time between heat and X-irradiation via kinetics similar to those for the removal of excess nuclear proteins; whereas the residual damage was reduced with time between X-irradiation and heat via kinetics similar to those of repair of radiation-induced DNA damage. Thus it appears that the excess nuclear proteins may play a major role in the inhibition of DNA strand break rejoining. However, an argument has been made that additional factors may be involved in the heat-induced inhibition of rejoining strand breaks (Jorritsma and Konings 1984).

(iii) Repair of UV-induced DNA damage. The repair of DNA damage following UV irradiation involves multiple sequential steps: (1) recognition of the damaged bases; (2) enzymatic removal of damaged pyrimidines or purines; (3) resynthesis of the excised strand; (4) ligation of the resynthesized DNA; and (5) reassembly of the repaired DNA into normal chromatin structure (Hanawalt *et al.* 1982). A study (Bodell *et al.* 1984) of the effects of hyperthermia (45 °C) on this process has shown the following. Repair

patches initiated prior to hyperthermia completed the reassembly step at the same rate as in unheated cells. However, the initiation of repair patches during hyperthermia was 50 to 70 per cent inhibited. Further, the reassembly into chromatin of repair patches synthesized during hyperthermia was inhibited 50 per cent relative to control rates of reassembly. However, ligation of the repair patch (and presumably its rate of resynthesis) proceeded at the control rate. Thus hyperthermia inhibits the excision repair process by interfering with at least two steps: e.g. initiation, either (1) or (2) and reassembly (5).

F. OVERVIEW

Having reviewed many of the heat effects on subcellular macromolecular systems, we considered the mechanisms of cell death. In the context of this question it is important to realize that most of the interphase cell death (measured in terms of cell lysis) resulting from hyperthermia in cultured cells occurs later than 24 hours post-hyperthermia (Zieklc-Temme and Hopwood 1982a, b; Leeper 1985). Therefore, the heat effects observed in cells immediately following heat exposures in the range of 43 °C to 47 °C are occurring in physiologically living cells, although a significant fraction are destined to reproductive cell death or interphase cell death. Another important factor to consider is that cells in different phases of the cell cycle may be dying by different mechanisms. For example, S and G_2 phase cells complete at least one mitotic event prior to cell death; whereas G_1 phase cells die prior to the completion of the first mitosis (Coss and Dewey 1983). The G_1 cells appear to progress through S to G_2 prior to cell death (Roti Roti et al. 1986). These considerations lead us to the following model to explain heat-induced cell killing.

Our model consists of five steps: (1) disruption of critical plasma membrane structure(s), presumably the plasma membrane-cytoskeleton attachment points; (2) collapse of the cytoskeleton toward the nucleus; (3) absorption of protein onto the nuclear matrix; (4) disruption of nuclear functions involving DNA supercoiling changes; (5) damage to critical nuclear structures, possibly DNA and/or structures involving DNA. This model represents a refinement of those proposed previously (Dewey et al. 1980; Roti Roti 1982; Waters and Roti Roti 1982; Wong and Dewey 1986).

Some experimental evidence for the model is presented in Figs 4 and 5. The results in Fig. 4 show that when HeLa cells are heated at 43 °C chronic thermotolerance develops but that excess nuclear protein content and cell survival remain correlated. The data in Fig. 5 show that this correlation holds following modification by two thermal sensitizers, ethanol and procaine, and two protectors, glycerol and acute thermotolerance. These data expand upon previously published work which shows that membrane active thermal sensitizing agents can modify heat effects on nuclear protein content (Roti Roti and Wilson 1984), DNA replication (Wong and Dewey 1982) and polymerase-β activity *in situ* (Spiro et al. 1982). These observations led to the notion that heat effects on the plasma membrane and in the nucleus are interrelated.

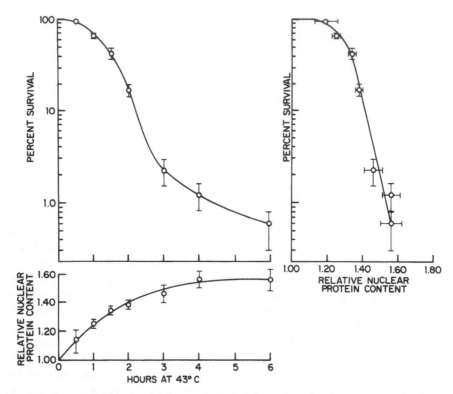

Figure 4: Correlation between nuclear protein content and cell survival during exposure to 43 °C. The surviving fraction is plotted as a function of heating time at 43 °C (upper left panel), while the relative nuclear protein content is plotted also as a function of heating time (lower left panel). Relative nuclear protein content was measured by FCM (Roti Roti *et al.* 1982). The right panel shows the surviving fraction plotted as a function of nuclear protein content (data replotted from the left panels). The plotted points represent the mean of three repeated experiments, while the bars represent ± one standard error.

These considerations form the basis of steps 1–3 in the model.

All five steps of the model are clearly present in S-phase cells. In such cells, step 5 of the model could involve the increased life-time of single stranded DNA regions (Wong and Dewey 1986). However, step 5 for cells in other cell-cycle stages is less clear. Thus, our model appears to be most reasonable for the killing of S-phase cells (Roti Roti 1982; Wong and Dewey 1986). However, critical information, namely the heat effects on cells in G_1 on the fidelity of DNA synthesis in the subsequent S-phase, is not available at this point in time. G_1 cells progress through S prior to interphase death in G_2 (Roti Roti *et al.* 1986). Another point is that FCM studies show that the relative change in nuclear protein content is the same (within the limits of the method) throughout the cell cycle. Further, these studies show that on an individual basis cells which survive show the same nuclear protein content as cells which die. Thus, it appears that the critical step in cell death depends upon what the cell is attempting to do while the excess nuclear proteins are present. This idea is supported by the observations that hyperthermia can restrict the DNA available for supercoiling changes (Roti Roti and Painter 1982; see Fig. 2) and that such changes are involved in many nuclear functions (Pardoll *et al.* 1980). These considerations prompted steps 3 to 5 in the model. Therefore, if nuclear protein changes are important in cell killing as the majority of evidence suggests, then all 5 steps in the model are necessary.

Experimental evidence demonstrating cause and effect relationships are available for portions of the model. For example, if nuclei are heated in the absence of cytostructure (i.e. isolated nuclei in the presence of whole cell

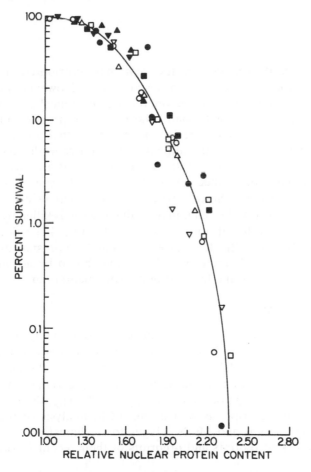

Figure 5: Cell survival as a function of nuclear protein content. The per cent surviving HeLa cells (ordinate) is plotted as a function of nuclear protein content (abscissa). Nuclear protein content was measured flow cytometrically (Roti Roti *et al.* 1982) immediately following heat exposure using aliquots of the same culture that was plated for single-cell for survival. Data from four separate heat-modification experiments were pooled; an experiment with ethanol (circles): an experiment with procaine (squares); an experiment with glycerol (triangles); and an experiment with thermotolerance (inverted triangles, first log only). In each case the heat alone (open symbols) controls were run in parallel with heat plus modifier (closed symbols). The heating temperature was 45 °C. The plotted symbols represent the means of at least three repeated experiments (error bars omitted for clarity).

sonicates), no change in nuclear protein is observed (Roti Roti and Winward 1980). Conversely, if the plasma membrane is damaged by detergent (Roti Roti and Winward 1980), alcohols or procaine (Roti Roti and Wilson 1984), then the nuclear protein content increases in a manner similar to that during hyperthermia. These results demonstrate steps 1 to 3 of the model. Steps 3 to 5 of the model remain to be demonstrated in a cause and effect manner for cell killing. However, steps 3 to 5 for DNA repair are demonstrated by the heat-induced inhibition of the excision of radiation-induced T' products (Warters and Roti Roti 1979). These results provide a partial demonstration of the model.

SUMMARY

The work produced over the past few years has contributed greatly to our picture of the changes occurring in a heated cell. However, many critical questions remain to be addressed before the definitive mechanisms of heat-induced cell killing can be established. The model we propose integrates the various heat-induced structural lesions in processes that leads to lesions in the genome which should be the ultimate target for reproductive cell death. However, given the extensive pleotropic effects of heat on cells, one can see that there are likely other mechanisms by which cells can be killed and/or damage reach the genome. Since hyperthermia affects molecules and molecular structures throughout the cell it may be that the cell is killed by summation of various lesions. Although this concept is attractive theoretically, it is an idea which will be very difficult to test experimentally. Our state of knowledge of the effects of hyperthermia on cell function should be advanced by the consideration of testable models and new experimental data.

ACKNOWLEDGEMENT

This work was supported by National Cancer Institute (NCI) Grant Numbers CA41102, 42591. The unpublished data shown here was collected with the support of NCI Grant CA29578, to Dr J.R. Stewart of the Department of Radiology, University of Utah. The authors would like to thank Ms Katherine McDonald for preparation of the manuscript; Dr R. Higashikubo for critical reading of the manuscript, as well as the FCM analysis of the presented data; Mr W.D. Wright and Ms N. Uygur for technical assistance; and all the people who sent reprints and preprints of their recent work.

REFERENCES

Alberts, B, Bray, D. *et al.* (1983). *Molecular Biology of the Cell.* Garland Publishing, New York.
Amalric, F., Simard, R., and Zalta, J.-P. (1969). Effet de la temperature supra-optimale sur les ribonucleoproteines et le RNA nucleolaire. II. Etude biochimique *Exptl. Cell Res.* **55**, 370–377.
Anderson, R.L., Minton, K.W., Li, G.C., and Hahn, G.M. (1981). Temperature-induced homeoviscous adaptation of Chinese hamster ovary cells. *Biochim. Biophys. Acta.* **641**, 334–348.

Anderson, R.L., Tao, T-W., and Hahn, G.M. (1984). Cholesterol: phospholipid ratios decrease in heat resistant variants of B16 melanoma cells. In *Hyperthermic Oncology,* Vol. 1 1984, pp. 123–126. Overgaard, J. (Ed.), Taylor and Francis, England.

Barrau, M.D., Blackburn, G.R., and Dewey, W.C. (1978). Effects of heat on the centrosomes of Chinese hamster ovary cells. *Cancer Res.* **38**, 2290–2294.

Bertazzoni, U., Stefanini, M., Noy, G.P., Giulotto, E., *et al.* (1976). Variations of DNA polymerases-α and -β during prolonged stimulation of human lymphocytes. *Proc. Natl. Acad. Sci. USA,* **73**, 785–789.

Blair, O.C., Winward, R.T., and Roti Roti, J.L. (1979). The effect of hyperthermia on the protein content of HeLa nuclei: A flow cytometric analysis. *Radiat. Res.* **78**, 474–484.

Bodell, W.J., Cleaver, J.E., and Roti Roti, J.L. (1984). Inhibition by hyperthermia of repair synthesis and chromatin reassembly of ultraviolet-induced damage to DNA. *Radiat. Res.* **100**, 87–95.

Bollum, F.J. (1975). Mammalian DNA polymerases. *Progr. Nucleic Acid Res. Mol. Biol.* **15**, 109–144.

Boonstra, J., Schamhart, D.H.J., DeLaat, S.W., and van Wijk, R. (1984). Analysis of K^+ and Na^+ transport and intracellular contents during and after heat shock and their role in protein synthesis in rat hepatoma cells. *Cancer Res.* **44**, 955–960.

Borrelli, M.J., Wong, R.S.L., and Dewey, W.C. (1986a). A direct correlation between hyperthermia-induced membrane blebbing and survival in synchronous G_1 CHO cells. *J. Cell. Physiol.* **126**, 181–190.

Borrelli, M.J., Carlini, W.G., Ransom, B.R., and Dewey, W.C. (1986b). Ion-sensitive microelectrode measurements of free intracellular chloride and potassium concentrations in hyperthermia-treated neuroblastoma cells. *J. Cell Physiol.* **129**, 175–184.

Brown, S.S., Malinoff, H.L., and Wicha, M.S. (1983). Connectin: A cell surface protein that binds both laminin and actin. *Proc. Natl. Acad. Sci. USA.* **80**, 5927–5930.

Buffa, P., Guarriera-Bobyleva, V., Muscatello, V., and Pasquali-Ronchetti, I. (1970). Conformational changes of mitochondria associated with uncoupling of oxidative phosphorylation *in vivo* and *in vitro. Nature (Lond).* **226**, 272–274.

Butler, A.P., Mardian, J.K.W., and Olins, D.E. (1985). Nonhistone chromosomal protein HMG 1 interactions with DNA. Fluorescence and thermal denaturation studies. *J. Biol. Chem.* **260**, 10613–10620.

Caizergues-Ferrer, M., Bouche, G., Banville, D., and Amalric, F. (1980). Effect of heat shock on RNA polymerase activities in Chinese hamster ovary cells. *Biochem. Biophys. Res. Commun.* **97**, 538–545.

Calderwood, S.K., and Hahn, G.M. (1983). Thermal sensitivity and resistance of insulin receptor binding. *Biochim. Biophys. Acta,* **756**, 1–8.

Cervera, J. (1978). Effects of thermic shock on HEp-2 cells. An ultrastructural and high-resolution autoradiographic study. *J. Ultrastruct. Res.* **63**, 51–63.

Cervera, M., Dreyfuss, G., and Penman, S. (1981). Messenger RNA is translated when associated with the cytoskeletal framework in normal and VSV-infected HeLa cells. *Cell* **23**, 113–120.

Chambon, P. (1978). The molecular biology of the eukaryotic genome is coming of age. *Cold Spring Harbor Symp. Quant. Biol.* **42**, 1209–1234.

Christiansen, E.N., and Kvamme, E. (1969). Effects of thermal treatment on mitochondria of brain, liver and ascites cells. *Acta Physiol. Scand.* **76**, 472–484.

Clark, E.P., and Lett, J.T. (1978). Possible mechanisms of hyperthermic inactivation of the rejoining of X-ray induced DNA strand breaks. In Streffer, C. (Ed.), *Cancer Therapy by Hyperthermia and Radiation.* Urban & Schwarzenberg, Munich, pp. 144–145.

Clark, E.P., Dewey, W.C., and Lett, J.T. (1981). Recovery of CHO cells from hyperthermic potentiation to X-rays: Repair of DNA and chromatin. *Radiat. Res.* **85**, 302–313.

Connor, W.G., Gerner, E.W., Miller, R.C. and Boone, M.L.M. (1977). Prospects for hyperthermia in human cancer therapy. Part II. Implications of biological and physical data for applications of hyperthermia to man. *Radiology* **123**, 497–503.

Corry, P.M., Robinson, S., and Getz, S. (1977). Hyperthermic effects on DNA repair mechanisms. *Radiology* **123**, 475–482.

Coss, R.A., Dewey, W.C., and Bamburg, J.R. (1979). Effects of hyperthermia (41.5 °) on Chinese hamster ovary cells analyzed in mitosis. *Cancer Res.* **39**, 1911–1918.

Coss, R.A., Dewey, W.C., and Bamburg, J.R. (1982). Effects of hyperthermia on dividing Chinese hamster ovary cells and on microtubules in vitro. *Cancer Res.* **42**, 1059–1071.

Coss, R.A., and Dewey, W.C. (1983). Mechanism of heat sensitization of G_1 and S-phase cells by procaine-HCl. *Radiation Research: Tumor Biology and Therapy.* pp. D6–05 Broerse,

J.J., Barendsen, G.W., Kal, H.B., and Van Der Kogel, A.J. (Ed.), Martinnus Nijhoff, Amsterdam.

Cress, A.E., Culver, P.S., Moon, T.E., and Gerner, E.W. (1982). Correlation between amounts of cellular membrane components and sensitivity to hyperthermia in a variety of mammalian cell lines in culture. *Cancer Res.* **42**, 1716–1721.

Cress, A.E., and Gerner, E.W. (1980). Cholesterol levels inversely reflect the thermal sensitivity of mammalian cells in culture. *Nature* **283**, 677–679.

Cress, A.E., and Bowden, G.T. (1983). Covalent DNA–protein crosslinking occurs after hyperthermia and radiation. *Radiat. Res.* **95**, 610–617.

Darzynkiewicz, Z., Traganos, F., Sharpless, T., and Melamed, M.R. (1975). Thermal denaturation of DNA *in situ* as studied by acridine orange staining and automated cytofluorometry. *Exptl. Cell Res.* **90**, 411–428.

Darzynkiewicz, Z., Traganos, F., Sharpless, Arlin, Z. and Melamed, M.R. (1976). Cytofluorometric studies on conformation of nucleic acids *in situ*. II. Denaturation of DNA. *J. Histochem. Cytochem.* **24**, 49–58.

Darzynkiewicz, Z., Traganos, F., Sharpless, T., and Melamed, M.R. (1977). Different sensitivity of DNA *in situ* in interphase and metaphase chromatin to heat denaturation. *J. Cell Biol.* **73**, 128–138.

Darzynkiewicz, Z., Traganos, F., Xue, S-B. and Melamed, M. (1981). Effect of n-butyrate on cell-cycle progression and *in situ* chromatin structure of L1210 cells. *Exp. Cell. Res.* **136**, 279–293.

DeBenedetti, A., and Baglioni, C. (1986). Activation of hemin-regulated initiation factor-2 kinase in heat-shocked HeLa cells. *J. Biol. Chem.* **261**, 338–342.

Defer, N., Kitzis, A., Kruh, J., Brahms, S. and Brahms, J. (1977). Effect of nonhistone proteins on thermal transition of chromatin and of DNA. *Nucl. Acids Res.* **7**, 2293–2306.

Dewey, W.C., Freeman, M.L., Raaphorst, G.P., Clark, E.P., Wong, R.S., Highfield, D.P., Spiro, J.S., Tomasovic, S.P., Denman, D.L., and Cross, R.A. (1980). Cell biology of hyperthermia and radiation. In *Radiation Biology in Cancer Research*, pp. 589–621. Meyn, R.E., and Withers, H.R. (Ed.), Raven Press, New York.

Dewey, W.C., Sapareto, S.A., and Betten, D.A. (1978). Hyperthermic radiosensitization of synchronous Chinese hamster cells: relationship between lethality and chromosomal aberrations. *Radiat. Res.* **76**, 48–59.

Dickson, J.A., and Calderwood, S.K. (1979). Effects of hyperglycemia and hyperthermia on the pH, glycolysis, and respiration of the Yoshida sarcoma *in vivo*. *J. Natl. Cancer Inst.* **63**, 1371–1375.

Dimitrov, S.I., Tsaneva, I.R., Pashev, I.G., and Markov, G.G. (1980). Structural rearrangement of histone/H1-depleted chromatin during thermal denaturation. *Biochem. Biophys. Acta* **610**, 392–399.

Dreskin, S.C., and Mayall, B.H. (1974). Deoxyribonucleic acid cytophotometry of stained human leukocytes. III. Thermal denaturation of chromatin. *J. Histochem. Cytochem.* **22**, 120–126.

Dube, D.K., Seal, G., and Loeb, L.A. (1977). Differential heat sensitivity of mammalian DNA polymerases. *Biochem. Biophys. Res. Commun.* **76**, 483–487.

Duncan, R., and Hershey, J.W.B. (1984). Heat shock-induced translational alterations in HeLa cells. Initiation factor modifications and the inhibition of translation. *J. Biol. Chem.* **259**, 11882–11889.

Falkner, F.G., Saumweber, H., and Biessman, H. (1981). Two *Drosophila melanogaster* proteins related to intermediate filament proteins of vertebrate cells. *J. Cell Biol.* **91**, 175–183.

Findly, R.C., and Pederson, T. (1981). Regulated transcription of the genes for actin and heat-shock proteins in cultured Drosophila cells. *J. Cell Biol.* **88**, 323–328.

Fuhr, J.E. (1974). Effect of hyperthermia on protein biosynthesis in L5178Y murine leukemic lymphoblasts. *J. Cell. Physiol.* **84**, 365–372.

Fuhr, J.E., Overton, M., and Leisy, M. (1974). Protective effect of cycloheximide upon protein synthesis by L5178Y cells exposed to hyperthermia. *Cytobios* **11**, 107–113.

Fuhr, J.E. and Overton, M. (1974). Effect of antinomycin D on protein synthesis in murine leukemia cells after hyperthermia. *Cytobios* **11**, 161–165.

Gerner, E.W., Cress, A.E., Stickney, D.G., Holmes, D.K., and Culver, P. (1980). Factors regulating membrane permeability alter thermal resistance. *Ann. N.Y. Acad. Sci.* **335**, 215–233.

Glass, J.R., DeWitt, R.G., and Cress, A.E. (1985). Rapid loss of stress fibers in Chinese hamster ovary cells after hyperthermia. *Cancer Res.* **45**, 258–262.

Goldstein, E.S., and Penman, S. (1973). Regulation of protein synthesis in mammalian cells. V. Further studies on the effect of actinomycin D on translation control in HeLa cells. *J. Mol. Biol.* **80**, 243–254.

Goldstein, E.S., Reichman, M.E., and Penman, S. (1974). The regulation of protein synthesis in mammalian cells. VI. Soluble and polyribosome associated components controlling *in vitro* polypeptide initiation in HeLa cells. *Proc. Natl. Acad. Sci. USA* **71**, 4752–4756.

Gonzalez-Mendez, R., Minton, K.W., and Hahn, G.M. (1982). Lack of correlation between membrane lipid composition and thermotolerance in Chinese hamster ovary cells. *Biochim. Biophys. Acta.* **692**, 168–170.

Guffy, M.M., Rosenberger, J.A., Simon, I., and Burns, C.P. (1982). Effect of cellular fatty acid alteration on hyperthermic sensitivity in cultured LK1210 murine leukemia cells. *Cancer Res.* **42**, 3625–3630.

Guttman, T., Vitek, A., and Pivec, L. (1977). High resolution thermal denaturation of mammalian DNAs. *Nucl. Acids Res.* **4**, 285–297.

Hahn, G.M., and Strande, D.P. (1976). Cytotoxic effects of hyperthermia and adriamycin on Chinese hamster cells. *J. Natl. Cancer Inst.* **57**, 1063–1067.

Hahn, G.M. (1982). *Hyperthermia and Cancer*, Plenum Press, N.Y.

Hanawalt, P.C., Cooper, P.K., Ganesan, A.K., Lloyd, R.S., Smith, C.A. and Zolan, M.E. (1982). Repair response to DNA damage: Enzymatic pathways in E.Coli and human cells. *J. Cell. Biochem.* **18**, 271–282.

Hayat, H., Brenner, H.J., and Friedberg, I. (1984). Hyperthermic cell membrane permeabilization in normal and transformed mouse fibroblasts. In *Hyperthermic Oncology*, Vol. I 1984 pp. 35–36. Overgaard, J. (Ed.), Taylor and Francis, London

Heacock, C.S., Brown, S.L., and Bamburg, J.R. (1982). *In vitro* inactivation of actin by heat. *Natl. Cancer Inst. Monogr.* **61**, 73–75.

Heine, U., Severak, L., Kondratick, J., and Bonar, R.A. (1971). The behavior of HeLa-S₃ cells under the influence of supranormal temperatures. *J. Ultrastruct. Res.* **34**, 375–396.

Henle, K.J., and Dethlefsen, L.A. (1978). Heat fractionation and thermotolerance. A review. *Cancer Res.* **38**, 1843–1851.

Henle, K.J., and Leeper, D.B. (1979). Effects of hyperthermia (45°) on macromolecular synthesis in Chinese hamster ovary cells. *Cancer Res.* **39**, 2665–2674.

Hidvegi, E.G., Yatvin, M.B., Dennis, W.H., and Hidvegi, A. (1980). Effect of altered membrane lipid composition and procaine on hyperthermic killing of ascites tumor cells. *Oncology (Basel)* **37**, 360–363.

Higashikubo, R., Uygur, N., and Roti Roti, J.L. (1986). Role of RNA and protein synthesis and turnover in the heat-induced increase in nuclear protein. *Radiat. Res.* **106** 278–282.

Howe, J.G., and Hershey, J.W.B. (1984). Translational initiation factor and ribosome association with the cytoskeletal framework fraction from HeLa cells. *Cell* **37**, 85–93.

Hume, S.P., and Field, S.B. (1977). Acid phosphatase activity following hyperthermia of mouse spleen and its implication in heat potentiation of X-ray damage. *Radiat. Res.* **72**, 145–153.

Hume, S.P., Rogers, M.A., and Field, S.B. (1978). Two qualitatively different effects of hyperthermia on acid phosphatase staining in mouse spleen, dependent on the severity of the treatment. *Int. J. Radiat. Biol.* **34**, 401–409.

Humphries, G.M.K., and Lovejoy, J.P. (1983). Dansyl lysine: A structure selective fluorescent membrane stain? *Biophys. J.* **42**, 307–310.

Hutchison, J.S., and Moldave, K. (1981). The effect of elevated temperature on protein synthesis in cell-free extracts of cultured Chinese hamster ovary cells. *Biochem. Biophys. Res. Commun.* **99**, 722–728.

Jorritsma, J.B.M., and Konings, A.W.T. (1984). The occurrence of DNA strand breaks after hyperthermic treatments of mammalian cells with and without radiation. *Radiat. Res.* **98**, 198–208.

Kampinga, H.H., Jorritsma, J.B.M., and Konings, A.W.T. (1985). Heat-induced alterations in DNA polymerase activity of HeLa cells and of isolated nuclei. Relation to cell survival. *Int. J. Radiat. Biol.* **47**, 29–40.

Klump, H. (1977). Thermodynamic values of the helix-coil transition of DNA in the presence of quaternary ammonium salt. *Biochim. Biophys. Acta* **475**, 605–610.

Klump, H., and Burkart, W. (1977). Calorimetric measurements of the transition enthalpy of DNA in aqueous urea solutions. *Biochim. Biophys. Acta* **475**, 601–604.

Konings, A.W.T. (1985). Development of thermotolerance in mouse fibroblast LM cells with modified membranes and after procaine treatment. *Cancer Res.* **45**, 2016–2019.

Konings, A.W.T., and Ruifrok, A.C.C. (1985). Role of membrane lipids and membrane fluidity in thermosensitivity and thermotolerance of mammalian cells. *Radiat. Res.* **102**, 86–98.

Korn, E.D. (1982). Actin polymerization and its regulation by proteins from non-muscle cells. *Physiol. Rev.* **62**, 672–737.

Kornberg, R.D. (1977). Structure of chromatin. *Ann. Rev. Biochem.* **46**, 931–954.

Kwock, L., Lin, P-S, Hefter, K., and Wallach, D.F.H. (1978). Impairment of Na^+-dependent amino acid transport in a cultured human T-cell line by hyperthermia and irradiation. *Cancer Res.* **38**, 83–87.

Lai, C-S, Hopwood, L.E., and Swartz, H.M. (1980). Electron spin resonance studies of changes in membrane fluidity of Chinese hamster ovary cells during the cell cycle. *Biochim. Biophys. Acta.* **602**, 117–126.

Leeper, D.B. (1985). Molecular and cellular mechanisms of hyperthermia alone or combined with other modalities. In *Hyperthermic Oncology* Vol. II, 1984, pp. 9–40. Overgaard, J. (Ed.) Taylor and Francis, England.

Lenk, R., Ransom, L., Kaulmann, Y., and Penman, S. (1977). A cytoskeletal structure with associated polyribosomes obtained from HeLa cells. *Cell* **10**, 67–78.

Lepock, J.R., Massicotte-Nolan, P., Rule, G.S., and Kruuv, J. (1981). Lack of a correlation between hyperthermic cell killing, thermotolerance, and membrane lipid fluidity. *Radiat. Res.* **87**, 300–313.

Lepock, J.R. (1982). Involvement of membranes in cellular responses to hyperthermia. *Radiat. Res.*, **92**, 433–438.

Lepock, J.R., Cheng, K-H., Al-Qysi, H., and Kruuv, J. (1983). Thermotropic lipid and protein transitions in Chinese hamster lung cell membranes: relationship to hyperthermic cell killing. *Can. J. Biochem. Cell Biol.* **61**, 421–427.

Lewis, P.N. (1977). A thermal denaturation study of chromatin and nuclease-produced chromatin fragments. *Can. J. Biochem.* **55**, 736–746.

Li, G.C., and Hahn, G.M. (1978). Ethanol-induced tolerance to heat and adriamycin. *Nature* **274**, 699–701.

Li, G.C., and Hahn, G.M. (1980). Adaptation to different growth temperatures modifies some mammalian cell survival responses. *Exp. Cell Res.***128**, 475–485.

Li, G.C., Shiu, E.C., and Hahn, G.M. (1980). Similarities in cellular inactivation by hyperthermia or by ethanol. *Radiat. Res.* **82**, 257–268.

Lin, P-S, Turi, Kwock, and Lu, R.C. (1982). Hyperthermic effect on microtubule organization. *Natl. Cancer Inst. Monogr.* **61**, 57–60.

Lin, P-S, Kwock, L., Hefter, K., and Wallach, D.F.H. (1978). Modification of rat thymocyte membrane properties by hyperthermia and ionizing radiation. *Int. J. Radiat. Biol.* **33**, 371–382.

Loeb, L.A. (1974). Eukaryotic DNA polymerases. In *The Enzymes*, Vol. 10 pp. 173–209. Boyer, P. (Ed.), Academic Press, New York.

Magun, B.E. (1981). Inhibition and recovery of macromolecular synthesis, membrane transport, and lysosomal function following exposure of cultured cells to hyperthermia. *Radiat. Res.* **87**, 657–669.

Magun, B.E., and Fennie, C.W. (1981). Effects of hyperthermia on binding, internalization, and degradation of epidermal growth factor. *Radiat. Res.* **86**, 133–146.

Mattern, M.R., and Painter, R.B. (1979). Dependence of mammalian DNA replication on DNA supercoiling. II. Effects of novobiocin on DNA synthesis in Chinese hamster ovary cells. *Biochim. Biophys. Acta.* **563**, 306–312.

Maul, G.G. (1982). *The Nuclear Envelope and the Nuclear Matrix.* Allen R. Liss, New York.

Mayrand, S., and Pederson, T. (1983). Heat shock alters nuclear ribonucleoprotein assembly in Drosophilia cells. *Molec. Cell. Biol.* **3**, 161–171.

McCormick, W., and Penman, S. (1969). Regulation of protein synthesis in HeLa cells: translation at elevated temperatures. *J. Mol. Biol.* **39**, 315–333.

Mehdi, S.Q., Recktenwald, D.J., Smith, L.M., Li, G.C., Armour, E.P., and Hahn, G.M. (1984). Effect of hyperthermia on murine cell surface histocompatibility antigens. *Cancer Res.* **44**, 3394–3397.

Mills, M.D., and Meyn, R.E. (1981). Effects of hyperthermia on repair of radiation-induced DNA strand breaks. *Radiat. Res.* **87**, 314–328.

Mills, M.D., and Meyn, R.E. (1983). Hyperthermic potentiation of unrejoined DNA strand breaks following irradiation. *Radiat. Res.* **95**, 327–338.

Mivechi, N.F., and Dewey, W.C. (1984). Effect of glycerol and low pH on heat-induced cell killing and loss of cellular DNA polymerase activities in Chinese hamster ovary cells. *Radiat. Res.* **99**, 352–362.

Mivechi, N.F., and Dewey, W.C. (1985). DNA polymerase α and β activities during the cell cycle and their role in heat radiosensitization in Chinese hamster ovary cells. *Radiat. Res.* **103**, 337–350.

Mondovi, B., Strom, R., Rotilio, G., Agro, F.A., Caviliere, R., and Rossi-Fanelli, A. (1969). The biochemical mechanism of selective heat sensitivity of cancer cells. I. Studies on cellular respiration. *Eur. J. Cancer* **5**, 129–136.

Mondovi, B., Agro, F.A., Rotilio, G., Strom, R., Moricca, G., and Rossi-Fanelli, A. (1969). The biochemical mechanism of selective heat sensitivity of cancer cells. II. Studies on nucleic acids and protein synthesis. *Eur. J. Cancer* **5**, 137–146.

Mulcahy, R.T., Gould, M.N., Hidvegi, E., Elson, C.E., and Yatvin, M.B. (1981). Hyperthermia and surface morphology of P388 ascites tumour M.B. cells: effects of membrane modifications. *Int. J. Radiat. Biol.* **39**, 95–106.

Ochoa, S. (1983). Regulation of protein synthesis initiation in eucaryotes. *Arch. Biochem. Biophys.* **223**, 325–349.

Oleinick, N.L. (1979). The initiation and elongation steps in protein synthesis: relative rates in Chinese hamster ovary cells during and after hyperthermic and hypothermic shocks. *J. Cell. Physiol.* **98**, 185–192.

Overgaard, K. and Overgaard, J. (1972a). Investigations on the possibility of a thermic tumour therapy. I. Short-wave treatment of a transplanted isologous mouse mammary carcinoma. *Eur. J. Cancer* **8**, 65–78.

Overgaard, K. and Overgaard, J. (1972b). Investigations on the possibility of thermic tumor therapy. II. Action of combined heat-roentgen treatment on a transplanted mouse mammary carcinoma. *Eur. J. Cancer.* **8**, 573–575.

Overgaard, J. (1976). Ultrastructure of a murine mammary carcinoma exposed to hyperthermia *in vivo. Cancer Res.* **36**, 983–995.

Panniers, R., and Henshaw, E.C. (1984). Mechanism of inhibition of polypeptide chain initiation in heat-shocked Ehrlich ascites tumour cells. *Eur. J. Biochem.* **140**, 209–214.

Panniers, Stewart, E.B., Merrick, W.C., and Henshaw, E.C. (1985). Mechanism of inhibition of polypeptide chain initiation in heat-shocked Ehrlich cells involves reduction of eukaryotic initiation factor 4F activity. *J. Biol. Chem.* **260**, 9648–9653.

Pardoll, D.M., Vogelstein, B., and Coffey, D.S. (1980). A fixed site of DNA replication in eucaryotic cells. *Cell* **19**, 527–538.

Pelham, H.R.B. (1984). Hsp 70 accelerates the recovery of nucleolar morphology after heat shock. *EMBO J.* **3**, 3095–3100.

Ray, B.K., Brendler, T.G., Adya, S, Daniels-McQueen, S., Kelvin-Miller, J., Hersey, J.W.B., Grifo, J.A., Merrick, W.C., and Thach, R.E. (1983). Role of mRNA competition in regulating translation: Further characterization of mRNA discriminatory initiation factors. *Proc. Natl. Acad. Sci. USA,* **80**, 663–667.

Reeves, O.R. (1972). Mechanisms of acquired resistance to acute heat shock in cultured mammalian cells. *J. Cell. Physiol.* **79**, 157–170.

Reichman, M., and Penman, S. (1973). Stimulation of polypeptide initiation *in vitro* after protein synthesis inhibition *in vivo* in HeLa cells. *Proc. Natl. Acad. Sci. USA* **70**, 2678–2682.

Reiter, T., and Penman, S. (1983). 'Prompt' heat shock proteins: translationally regulated synthesis of new proteins associated with the nuclear matrix-intermediate filaments as an early response to heat shock. *Proc. Natl. Acad. Sci. USA* **80**, 4737–4741.

Rice, G.C., Fisher, G., Devlin, M., Humphries, G.M.K., and Qasim Mehdi, S. (1985a). Use of N-ε-dansyl-L-lysine and flow cytometry to identify heat-killed mammalian cells. *Int. J. Hyperthermia* **1**, 185–191.

Rice, G.C., Gray, J.W., and Dewey, W.C. (1985b). FACS analysis of a hyperthermia-induced alteration in Hoechst 33342 permeability and direct measurement of its relationship to cell survival. *J. Cell. Physiol.* **122**, 387–396.

Roth, D. (1978). Fluorescent probe study of DNA conformation in briefly heated human squamous cells: *J. Natl. Cancer Inst.* **60**, 97–99.

Roti Roti, J.L., and Winward, R.T. (1978). The effects of hyperthermia on the protein to DNA ratio of isolated HeLa chromatin. *Radiat. Res.* **74**, 159–169.

Roti Roti, J.L., Henle, K.J., and Winward, R.T. (1979). The kinetics of increase in chromatin protein content in heated cells: A possible role in cell killing. *Radiat. Res.* **78**, 522–531.

Roti Roti, J.L., and Winward, R.T. (1980). Factors affecting the heat-induced increase in protein content of chromatin. *Radiat. Res.* **81**, 138–144.

Roti Roti, J.L., Higashikubo, R., Blair, O.C., and Uygur, N. (1982). Cell-cycle position and nuclear protein content. *Cytometry* **3**, 91–96.

Roti Roti, J.L., and Painter R.B. (1982). Effects of hyperthermia on the sedimentation of nucleoids from HeLa cells in sucrose gradients. *Radiat. Res.* **89**, 166–175.

Roti, Roti, J.L. (1982). Heat-induced cell death and radiosensitization: Molecular Mechanisms. *Natl. Cancer Inst. Monogr.* **61**, 3–10.

Roti Roti, J.L., Higashikubo, R., and Mace, M. (1984). Protein cross-migration during isolation of nuclei from mixtures of heated and unheated HeLa cells. *Radiat. Res.* **98**, 107–114.

Roti Roti, J.L., and Wilson, C.F. (1984). The effects of alcohols, procaine and hyperthermia on the protein content of nuclei and chromatin. *Int. J. Radiat. Biol.* **46**, 25–33.

Roti Roti, J.L., Wright, W.D., Higashikubo, R., and Dethlefsen, L.A. (1985). DNase I sensitivity of nuclear DNA measured by flow cytometry. *Cytometry* **6**, 191–208.

Roti Roti, J.L., Uygur, N., and Higashikubo, R. (1986). Nuclear protein following heat shock: Protein removal kinetics and cell cycle rearrangements. *Radiat. Res.* **107**, 250–261.

Roti Roti, J.L., and Wright, W.D. (1987). Visualization of DNA loops in nucleoid from HeLa cells; assays for DNA damage and repair. *Cytometry* **8**, 461–467.

Ruifrok, A.C.C., Kanon, B., and Konings, A.W.T. (1985a). Correlation between cellular survival and potassium loss in mouse fibroblasts after hyperthermia alone and after a combined treatment with X-rays. *Radiat. Res.* **101**, 326–331.

Ruifrok, A.C.C., Kanon, B., and Konings, A.W.T. (1985b). Correlation of colony forming ability of mammalian cells with potassium content after hyperthermia under different experimental conditions. *Radiat. Res.* **103**, 452–454.

Schlesinger, M.J., Ashburner, M., and Tissieres, A. (1982). In *Heat Shock from Bacteria to Man.* Cold Spring Harbor Lab Press, Cold Spring Harbor, NY.

Schliwa, M. (1981). Proteins associated with cytoplasmic actin. *Cell,* **25**, 587–590.

Schneider, E., Muller, B., and Schindler, R. (1985). Control of DNA polymerase α, β, and γ activities in heat and cold-sensitive mammalian cell-cycle mutants. *Biochim. Biophys. Acta* **825**, 375–383.

Schochetman, G., and Perry, R.P. (1972). Characterization of the messenger RNA released from L cell polyribosomes as a result of temperature shock. *J. Mol. Biol.* **63**. 577–590.

Seale, R.L. (1975). Assembly of DNA and protein during replication in HeLa cells. *Nature* **255**, 247–249.

Seligy, V.L., and Poon, N.H. (1978). Alteration in nucleosome structure induced by thermal denaturation. *Nucl. Acids Res.* **5**, 2233–2252.

Simard, R., Amalric, F., and Zalta, J.-P. (1969). Effet de la temperature supra-optimale sur les ribonucleoproteines et le RNA nucleolaire. I Etude ultrastructural *Exptl. Cell Res.* **55**, 359–369.

Simard, R., and Bernhard, W. (1967). A heat-sensitive cellular function located in the nucleolus. *J. Cell Biol.* **34**, 61–76.

Slusser, H., Hopwood, L.E., and Kapiszewska, M. (1982). Inhibition of membrane transport by hyperthermia. *Natl. Inst. Cancer Monogr.* **61**, 85–87.

Souliman, S., and Chapman, I.V. (1981). Diffusion kinetics of tritiated uridine into HeLa cells previously exposed to hyperthermia. *Int. J. Radiat. Biol.* **39**, 31–38.

Spiro, I.J., Denman, D.L., and Dewey, W.C. (1982). Effect of hyperthermia on CHO DNA polymerases α and β. *Radiat. Res.* **89**, 134–149.

Spiro, I.J., Denman, D.L., and Dewey, W.C. (1983). Effect of hyperthermia on isolated DNA β polymerase. *Radiat. Res.* **95**, 68–77.

Stevenson, A.P., Galey, W.R., and Tobey, R.A. (1983). Hyperthermia-induced increase in potassium transport in Chinese hamster cells. *J. Cell. Physiol.* **115**, 75–86.

Streffer, C. (1985). Mechanism of heat injury. In *Hyperthermic Oncology* Vol. II, 1984, pp. 213–222. Overgaard, J. (Ed)., Taylor and Francis, London.

Strom, R., Santoro, S., Crifo, C., Bozzi, A., Mondovi, B., and Rossi-Fanelli, A. (1973). The biochemical mechanism of selective heat sensitivity of cancer cells. IV. Inhibition of RNA synthesis. *Eur. J. Cancer* **9**, 103–112.

Strom, R., Crifo, C., Rossi-Fanelli, A., and Mondovi, B. (1977). Biochemical aspects of heat sensitivity of tumour cell. In *Selective Heat Sensitivity of Cancer Cells*, pp. 7–35, Rossi-Fanelli, A., Cavaliere, R., Mondovi, B., and Morricca, G. (Eds), Springer Verlag, Berlin, Heidelberg, New York.

Subjeck, J.R., Shyy, T., Shen, J., and Johnson, R.J. (1983). Association between the mammalian 110,000-dalton heat-shock protein and nucleoli. *J. Cell Biol.* **97**, 1389–1395.

Thomas, G.P., Welch, W.J., Mathews, M.B., and Feramisco, J.R. (1982). Molecular and cellular effects of heat shock and related treatments of mammalian tissue-culture cells. *Cold Spring Harbor Symposium on Quantitative Biology* **46**, 985–996.

Tomasovic, S.P., Turner, G.N., and Dewey, W.G. (1978). Effects of hyperthermia on nonhistone proteins isolated with DNA. *Radiat. Res.* **73**, 535–552.

Turano, G., Ferraro, A., Strom, R., Cavaliere, R., and Rossi-Fanelli, A. (1970). The biochemical mechanism of selective heat sensitivity of cancer cells. III. Studies on lysosomes. *Eur. J. Cancer.* **6**, 67–72.

Van Bergen En Henegousen, P.M.P., Jordi, W.J.R.M.,Van Dongen, G., Ramaekers, F.C.S., Amesz, H., and Linneman, W.A.M. (1985). Studies on a possible relationship between alterations in the cytoskeleton and induction of heat shock protein synthesis in mammalian cells. *Int. J. Hyperthermia* **1**, 69–83.

Van Venrooij, W.J., Sillekens, P.T.G., Van Eekelen, C.A.G., and Reinders, R.J. (1981). On the association of mRNA with the cytoskeleton in uninfected and adenovirus-infected human KB cells. *Exptl. Cell Res.* **135**, 79–91.

Vidair, C.A., and Dewey, W.C. (1986). Evaluation of a role for intracellular Na^+, K^+, Ca^{2+}, and Mg^{2+} in hyperthermic cell killing. *Radiat. Res.* **105**, 187–200.

Vogelstein, B., Pardoll, D.M., and Coffey, D.S. (1980). Supercoiled loops and eucaryotic DNA replication. *Cell* **22**, 79–85.

Warocquier, R., and Scherrer, K. (1969). RNA metabolism in mammalian cells at elevated temperature. *Eur. J. Biochem.* **10**, 362–370.

Warters, R.L., and Roti Roti, J.L. (1979). Excision of X-ray-induced thymine damage in chromatin from heated cells. *Radiat. Res.* **79**, 113–121.

Warters, R.L., Roti Roti, J.L., and Winward, R.T. (1980). Nucleosome structure in chromatin from heated cells. *Radiat. Res.* **84**, 504–513.

Warters, R.L., and Roti Roti, J.L. (1981). The effect of hyperthermia on replicating chromatin. *Radiat. Res.* **88**, 69–78.

Warters, R.L. (1982). DNA damage in heated cells. *Natl. Cancer Inst. Monogr.* **61**, 45–47.

Warters, R.L., and Roti Roti, J.L. (1982). Hyperthermia and the cell nucleus. *Radiat. Res.* **92**, 458–462.

Warters, R.L., and Henle, K.J. (1982). DNA degradation in heated CHO cells. *Cancer Res.* **42**, 4427–4432.

Warters, R.L., and Stone, O.L. (1983a). The effects of hyperthermia on DNA replication in HeLa cells. *Radiat. Res.* **93**, 71–84.

Warters, R.L., and Stone, O.L. (1983b). Macromolecule synthesis in HeLa cells after thermal shock. *Radiat. Res.* **96**, 646–652.

Warters, R.L., and Stone, O.L. (1984a). The sedimentation coefficient and buoyant density of nucleosomes from replicating chromatin in heated cells. *Radiat. Res.* **98**, 354–361.

Warters, R.L., and Stone, O.L. (1984b). Histone protein and DNA synthesis in HeLa cells after thermal shock. *J. Cell. Physiol.* **118**, 153–160.

Warters, R.L., Brizgys, L.M., Axtell-Bartlett, J. (1985). DNA damage production in CHO cells at elevated temperatures. *J. Cell. Physiol.* **124**, 481–486.

Warters, R.L., Yasui, L.S., Sharma, R., and Roti Roti, J.L. (1986). Heat shock (45 °C) results in an increase of nuclear matrix protein mass in HeLa cells. *Int. J. Radiat. Biol.* **50**, 253–268.

Weintraub, H. (1976). Cooperative alignment of nu bodies during chromosome replication in the presence of cycloheximide. *Cell* **9**, 419–422.

Weissbach, A. (1975). Vertebrate DNA polymerases. *Cell* **5**, 101–108.

Welch, W.J., and Feramisco, J.R. (1984). Nuclear and nucleolar localization of the 72000-dalton heat shock protein in heat-shocked mammalian cells. *J. Biol. Chem.* **259**, 4501–4513.

Welch, W.J., and Suhan, J.P. (1985). Morphological study of the mammalian stress response: Characterization of changes in cytoplasmic organelles, cytoskeleton and nucleoli, and appearance of intranuclear actin filaments in rat fibroblasts after heat-shock treatment. *J. Cell. Biol.* **101**, 1198–1211.

Westermark, N. (1927). The effect of heat upon rat tumors. *Skand. Arch. Physiol.* **52**, 257–302.

Westra, A., and Dewey, W.C. (1971). Variation in sensitivity to heat shock during the cell-cycle of Chinese hamster cells *in vitro. Int. J. Radiat. Biol.* **19**, 467–477.

Wheeler, K.T., and Warters, R.L. (1982). Influence of heat on the chromatin structure in HeLa nuclei. *Radiat. Res.* **90**, 204–215.

Wong, R.L., and Dewey, W.C. (1982). Molecular studies on the hyperthermic inhibition of DNA synthesis in Chinese hamster ovary cells. *Radiat. Res.* **92**, 370–395.

Wong, R.S.L., and Dewey, W.C. (1986). Effect of hyperthermia on DNA synthesis. In *Hyperthermia in Cancer Treatment*, pp. 80–91. Anghileri, L.J., and Robert, J. (Eds), CRC Press, Inc., Boca Raton, FL.

Wright, W.D., Higashikubo, R., Roti Roti, J.L. (1988). Fluorescent methods for studying subnuclear particles. In *Methods in Flow Cytometry* Darzynkiewicz, Z. (Ed.). In press.

Yatvin, M.B. (1977). The influence of membrane lipid composition and procaine on hyperthermic death of cell. *Int. J. Radiat. Biol.* **32**, 513–521.

Yau, P., Thorne, A.W., Imai, B.S., Matthews, H.R., and Bradbury, E.M. (1982). Thermal denaturation studies of acetylated nucleosomes and oligunucleosomes. *Eur. J. Biochem.* **129**, 281–288.

Yi, P.N. (1979). Cellular ion content changes during and after hyperthermia. *Biochem. Biophys. Res. Commun.* **91**, 177–181.

Yi, P.N. (1983). Hyperthermia-induced intracellular ionic level changes in tumor cells. *Radiat. Res.* **93**, 534–544.

Zielke-Temme, B., and Hopwood, L. (1982a). Time-lapse cinemicrographic observations of heated G1-phase Chinese hamster ovary cells. I. Division probabilities and generation times. *Radiat. Res.* **92**, 320–331.

Zielke-Temme, B., and Hopwood, L. (1982b). Time-lapse cinemicrographic observations of heated G1-phase Chinese hamster ovary cells. II. Cell death, fusion, and multipolar divisions. *Radiat. Res.* **92**, 332–342.

Hyperthermia and Oncology, Vol. 1, pp. 57–82 (1988)
Urano and Douple (Eds.)
© 1988 VSP.

Chapter 3

Response of cultured mammalian cells to hyperthermia

KURT J. HENLE[1] and JOSEPH L. ROTI ROTI[2]
[1]*University of Arkansas for Medical Sciences, Department of Medicine, Division of Hematology/Oncology, Little Rock, AR 72205, USA, and* [2]*Washington University School of Medicine, Mallinckrodt Institute of Radiology, Radiation Oncology Center, Section of Cancer Biology, St. Louis, MO 63108 USA*

A. MANIFESTATION OF CELLULAR HEAT DAMAGE

Hyperthermia has profound effects on all aspects of cellular biochemistry, morphology and function. The response of cultured cells to thermal stress has been the subject of hundreds of studies, some of which date back to the early 1970s. In spite of extensive accumulation of observations and data, however, our understanding of mechanistic relationships between events that lead from thermal damage to cell death has advanced very little. Reviews of thermal effects on mammalian cells, therefore, remain essentially phenomenological summaries of reported observations. This chapter will review the response of complete cells to hyperthermia; a parallel review of heat effects on individual cellular organelles is included within a separate chapter of this volume (Roti Roti and Laszlo, Chapter 2). The latter review also covers possible molecular mechanisms of cell killing.

Hyperthermia induces a host of alterations that can be studied in complete cells. The majority of these lesions correlate with heat 'dose' and with heat-induced cell killing. The most commonly accepted endpoint of irreversible heat damage is the loss of reproductive survival, i.e., the loss of colony forming ability. The scoring of reproductive survival is based on a criterion of 50 or more cells per colony following an incubation period of 1 to 2 weeks. Although this criterion was based on measurements of division probabilities following irradiation, the same criterion was also adopted for heat survival measurements without much additional experimental support.

Many other endpoints can correlate with cell survival after hyperthermia. One of the most notable effects in heated cells is the appearance of membrane blebbing (Borrelli *et al.* 1986); as a result, membrane lesions were suspected for a long time to be the cause of cellular heat death (Heilbrunn 1954; Mondovi *et al.* 1969; Bowler *et al.* 1973; Strom *et al.* 1973). However, Strom

(1973) pointed out that many morphological alterations of membranes from heated cells were quite reversible and, thus, might not participate in the sequence of events leading to heat-induced loss of reproductive survival. Scientific opinions continue to cycle between sentiments that attribute a key role to the plasma membrane in cellular heat death (Hahn 1982) and those that reduce membrane lesions to one of many readily reversible heat lesions. For example, a recent study of heat-induced membrane effects measured the efflux of cellular potassium, failure of the plasma membrane to exclude trypan blue, and eventual cell lysis (Ruifrock *et al.* 1987). All of those endpoints correlated, in general, with the extent of heat damage and with time after heating, but could be dissociated from reproductive survival by manipulating the serum content of the medium. Furthermore, three different cell lines exhibited variable degrees of membrane heat damage, as determined by these endpoints, even though they had been exposed to isotoxic heat 'doses'.

More subtle, but equally irreversible effects of heat damage can be demonstrated in the elevation of sister chromosome exchanges, chromosome aberrations in the S-phase and oncogenic transformations; genotoxic effects of hyperthermia and the potential role of DNA damage in cellular heat death are reviewed separately elsewhere (Nagle 1987). The majority of cellular heat effects in this review, therefore, are based on measurements of reproductive survival.

B.   RELATIVE HEAT SENSITIVITIES OF VARIOUS CELL LINES

The most commonly cited rationale for the use of hyperthermia in the early literature of hyperthermia in cancer treatment was the putative differential heat sensitivity of tumors, based on a hypothetical intrinsic characteristic of transformed cells (reviewed in Cavaliere *et al.* 1967; Giovanelli and Mondovi 1977; Hahn 1982). This sweeping generalization proved indefensible; instead, differential heat sensitivity today is generally accepted to be a consequence of solid tumor physiology. Poorly developed blood flow, relative nutrient deprivation and the accumulation of metabolic waste products represent the current dogma underlying the remarkable clinical response of spontaneous and transplanted tumors to hyperthermia (Chapters 6, 7, this volume).

1.   Established cell lines
Is it possible to predict the heat sensitivity of specific cell lines, based on their tissue of origin? In general, the answer is no. Figure 1 exemplifies the variability in cellular heat sensitivity for a number of established lines that were derived either from different species (Fig. 1A; Raaphorst *et al.* 1979) or even from the same type of tissue within the same species (Fig. 1B; Rofstad and Brustad 1984). The data illustrate a number of principles.

(a) Cellular heat sensitivity can vary enormously between specific mammalian cell lines even though cells are grown and heated under the same culture conditions. Cell survival after 5 hours at 42.5 °C differed by more than four decades between mouse LP59 and pig kidney CCL33 cells.

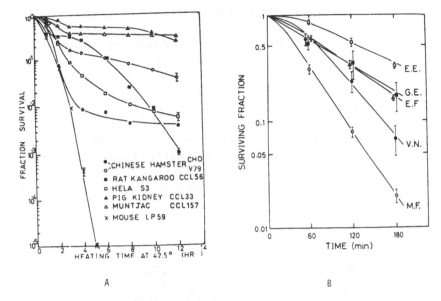

A   B

Figure 1: Comparative hyperthermia survival curves at 42.5 °C. Panel A: Cell lines derived
from a variety of sources, as specified in the graph, were subcultured 4 to 5 hours prior to
hyperthermia and heated under identical culture conditions in F12 medium, supplemented with
10 per cent fetal bovine serum and antibiotics. Plating efficiencies for unheated control cells
ranged from 32 per cent (pig kidney cells) to 86 per cent (Chinese hamster ovary cells). Cell
survival was calculated based on the number of colonies formed in heated vs unheated control
flasks (Raaphorst *et al.* 1979; reprinted with permission). Panel B: Five different human melanoma
cell lines were derived from histologically confirmed metastatic tumor tissue and transplanted
directly into athymic mice without prior adaptation in tissue culture. Tumors were serially
transplanted subcutaneously to the flank; passage 32 to 55 was used for measuring the heat
response in culture. Single cell suspensions were prepared from tumor tissue and cells were
plated in soft agar. After the experiment, tubes containing cells in agar were incubated 3 to 6
weeks at 37 °C with weekly replacement of medium. The plating efficiency of unheated cells
ranged from 5 per cent to 80 per cent (Rofstad and Brustad 1984; reprinted with permission).

(b) Cell killing at 42.5 °C was modulated in some, but not in all cell lines
during heating by the development of thermotolerance. Thermotolerance
appeared 3 to 5 hours after continuous heating at 42.5 °C; the two exceptions
in Figure 1A are represented by mouse LP59 cells and the marsupial line
CCL56. The capacity for thermotolerance development did not correlate
with initial heat sensitivity or other identifiable parameters.

(c) Two cell lines, both of which were derived from Chinese hamster tissues
(V79: lung; CHO: ovary), exhibited similar thermotolerance kinetics and
magnitude, but differed in terms of absolute survival by more than one
decade following hyperthermia of 42.5 °C for more than 4 hours.

(d) Cell lines derived from the same tissue, i.e. human melanoma xenografts,
but from different individuals (Fig. 1B), can span a range of heat sensitivities
comparable to that observed between different species shown in Figure 1A.
However, heating times in Fig. 1B extended only to 3 hours which is too
short for determining whether or not thermotolerance development varied
between the melanoma sublines. No correlations were found between the *in*

vitro and the *in vivo* heat sensitivity of these human xenografts (Rofstad and Brustad 1984).

Figure 2 shows the heat response of the same cell lines used in Fig. 1A, but at 45.5 °C instead of 42.5 °C (Raaphorst *et al.* 1979). At the higher temperature none of the cells developed thermotolerance during heating; the resulting survival curves, therefore, are less complex. Pig kidney cells remained the most heat-resistant cells, but the Chinese hamster cell lines appeared just as sensitive as the mouse LP59 cells at 45.5 °C. Muntjac (Indian deer) and pig kidney cells which showed similar heat sensitivities at 42.5 °C exhibited more than three decades of differential cell killing after 80 minutes at 45.5 °C. Thus, differential heat sensitivity of two cell lines at one specific temperature may not be applicable for predicting their relative heat sensitivity to other temperatures.

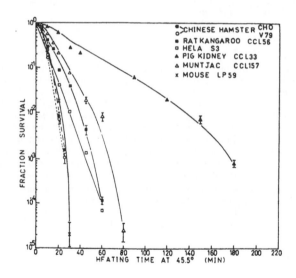

Figure 2: Comparative cell survival curves at 45.5 °C. Parallel experiments with cell lines in Fig. 1A were performed at 45.5 °C using identical experimental conditions. Relative heat sensitivities at 42 °C and 45.5 °C were not identical (Raaphorst *et al.* 1979; reprinted with permission).

2.   Endothelial cells

In spite of the complex time–temperature relationships and the heterogeneity of the cellular heat response, illustrated in Figs 1 and 2, a limited number of generalizations may be permissible. These generalizations refer to cells derived from those tissues that may limit therapeutic applications of hyperthermia or irradiation. For example, the tolerance of capillary endothelial cells has been suggested as a limiting factor for radiation tolerance of normal tissues. Cloned endothelial cells, derived from bovine aorta, however, suggest

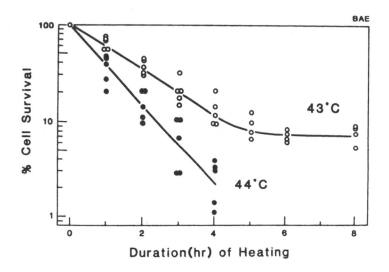

Figure 3: Hyperthermia survival curves of exponentially growing bovine aorta endothelial (BAE) cells. Endothelial cells, passage 7 to 15 (weekly 1:10 subculture), were grown in DMEM with 10 per cent calf serum and conditioned medium from cultures of mouse sarcoma 180 cells. Plating efficiency was approximately 11 per cent. BAE cells were heated 16 hours after plating and colonies were scored after 8 days at 37 °C. Symbols represent individual survival values from replicate experiments (Rhee and Song 1984; reprinted with permission).

that endothelial cells may not be treatment-limiting for hyperthermia; cell survival curves 43 ° and 44 °C (Fig. 3, Rhee and Song 1984) suggest relatively high heat resistance. After hyperthermia of 8 hours at 43 °C, cell survival was comparable to that of heat-resistant pig kidney or Muntjac cells that were heated over the same period at the lower temperature of 42.5 °C (Fig. 1A).

Ordinarily, endothelial cells do not form colonies and reproductive survival assays may reflect heat sensitivities of particular subpopulations (Fajardo *et al*. 1985). As an alternative to colony forming ability, Fajardo *et al*. (1985) used the cellular attachment capacity and cellular multiplicity in lieu of colony formation as a measure of heat sensitivity. With these endpoints they showed that both human and murine capillary endothelial cells were more heat sensitive than human foreskin fibroblasts and that heat sensitivity increased slightly when endothelial cells were stimulated with endothelial cell growth factor prior to hyperthermia. Heat sensitivity scored by the criterion of attachment to substrate was similar to the data in Fig. 3: for example, 30 minutes at 45 °C yielded 'survival' values of 40–50 per cent, dependent on the time of scoring at 1 or 3 days post-hyperthermia. The effect of hyperthermia on angiogenesis, however, is significantly more heat sensitive (Fajardo *et al*. 1987). For example, 50 per cent inhibition of angiogenesis required only 30 minutes at 42 °C, possibly reflecting the elevated heat sensitivity of growth-stimulated endothelial cells in the neoplastic vasculature, coupled with long heat-induced growth delays.

3. Intestinal cells

In contrast to the relative heat resistance of normal endothelial cells, normal bone marrow cells (Bromer et al. 1982; Kalland and Dahlquist 1983; Van Zant et al. 1983; Blackburn et al. 1984) and the intestinal crypt/villus system (reviewed by Hume 1985) are quite heat sensitive. These same cell systems are dose-limiting both for whole-body irradiation and for whole-body hyperthermia. Heat effects to the intestinal crypt/villus system is mediated by cellular effects and loss of intestinal function (Hume 1985). Heat effects could also be mediated indirectly by an enhanced susceptibility to systemic sepsis (Henle 1982). The intestinal crypt/villus has not been studied in vitro and the dependence of crypt damage on its proximity to the mesenteric blood flow has been ascribed variously to relative nutrient deprivation during hyperthermia (Henle 1982) or to cooling of crypts near the mesentery by circulating blood (Hume 1985).

4. Bone marrow cells

Cells derived from the bone marrow, as a general rule, are quite heat sensitive. This sensitivity is illustrated in Fig. 4 with survival data for human bone marrow cells, measured as a function of temperature (Bromer 1982). Comparison of Fig. 4A with Fig. 1 at 42.5 °C, indicates that the heat sensitivity for human marrow and mouse LP59 cells was similar and approached that of the most heat-sensitive human melanoma line, designated by M.F. (in Fig. 1). Furthermore, Fig. 4B indicates that human marrow cells, like mouse LP59 cells, did not develop thermotolerance during heating at 42.5 °C.

Cellular heat sensitivity is highly variable not only between cells from homologous tissues taken from different individuals and between different tissues from the same individual, but also between homologous tissues derived from different species. Thus, murine marrow cells (Kobayashi 1985) appear significantly more heat sensitive than human marrow cells (Bromer 1982; Blackburn 1984). Hyperthermia of 30 minutes, 44 °C, for example, reduced survival to 1 per cent for murine CFU-c in either normal or regenerating marrow (Kobayashi 1985). However, the same heat treatment reduced survival to less than 30 per cent in human marrow cells (Fig. 4A).

Within a population of hematopoietic cells erythroid precursors are more heat sensitive than precursors of macrophages, granulocytes and megakaryocytes (Van Zant 1983). Furthermore, natural killer cells are more heat sensitive than other lymphoid cells (Kalland 1983). These cells are believed to participate in immunological antitumor effects and may be inhibited following whole body hyperthermia.

C. RELATIONSHIP BETWEEN CELLULAR SENSITIVITY TO HYPER-THERMIA AND RADIATION

The potential relationship between cellular heat sensitivity and sensitivity to X-rays or ultraviolet radiation was explored by a number of investigators (Joshi et al. 1977; Gerweck and Burlett 1978; Raaphorst et al. 1979). The

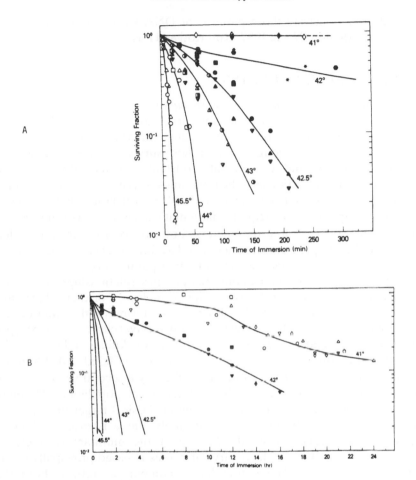

Figure 4: Survival curves of human bone marrow cells between 41 °C and 45.5 °C. Human bone marrow aspirates were separated from RBCs and plated in T25 culture flasks with McCoy's medium supplemented with 10 per cent fetal bovine serum, 8 per cent human AB plasma, 4 per cent human placental conditioned medium, heparin and antibiotics in soft agar 3 hours, 37 °C prior to experiments. After experiments cells were incubated 7 to 12 days before scoring colonies (CFU-c). The plating efficiency ranged between 0.01 per cent to 0.05 per cent. Each symbol represents a separate experiment, using different individuals on different days. Panel A: Effects of acute hyperthermia; Panel B: Effects of chronic hyperthermia (Bromer *et al*. 1982; reprinted with permission).

data from these studies showed that neither cellular sensitivity to U.V., nor to X-rays correlated with cellular heat sensitivity. These findings suggested that mechanisms of heat killing are not the same as those that kill cells by ionizing radiation. This conclusion was reinforced by other studies that showed radiation sensitizers, like bromodeoxyuridine or iododeoxyuridine, incorporated into cellular DNA, did not affect cellular heat sensitivity (Raaphorst *et al*. 1984). Furthermore, fibroblasts from normal controls or from radiation-sensitivity human subjects with ataxia telangiectasia had similar heat sensitivities (Raaphorst and Azzam 1982). Studies of heat-induced DNA damage also failed to demonstrate those patterns of DNA strand breaks that result from ionizing radiation (Nagle 1987; Roti Roti and Laszlo 1987 and Chapter 2).

D. KINETICS OF HEAT KILLING

1. Interpretations of cellular heat survival curves

Figures 1 to 4 show families of survival curves for specific cell lines. The survival curves represent the relatively narrow temperature range of approximately 41 °C to 46 °C. Although some of these survival curves are described by first order kinetics and a single rate constant, the majority of survival curves are more complex. For short heating times, survival curves may display a 'shoulder'; for extended heating periods of 3 to 6 hours many survival curves became biphasic. An adequate mathematical description of such curves requires at least two or three parameters. For any specific survival model, one can describe temperature-dependent cell killing in terms of the temperature dependence of appropriate survival curve parameters.

The most common description of time-temperature relationships for heat-induced cell killing is based on the Arrhenius formalism. For these Arrhenius analyses, cell survival is generally described in terms of the multitarget, single-hit survival model. This model represents survival curves by an extrapolation number, n, and an exponential rate of cell killing, related to the reciprocal of the dose required per lethal event (D_0) (Elkind and Whitmore 1967). Biphasic survival curves can be described by a second value of D_0 to represent the heat sensitivity of the thermotolerant cell population. Although these parameters have a conceptual meaning in classical target theory, their application to thermal biology does not imply the existence of discrete subcellular heat 'targets'. Instead, the use of the multitarget, single-hit formalism for parameterizing heat-survival curves is based principally on its descriptive convenience. The extrapolation number, n, reflects both the cellular capacity to accumulate and/or repair heat damage. In some cases, the temperature rise in the culture flask during the transition from 37 °C to a final hyperthermic temperature also is included in n.

Figure 5 shows a typical Arrhenius plot for the two survival-curve parameters, n and D_0, based on data from a number of established cell lines (Henle and Dethlefsen 1980). The data were corrected for temperature transients during the heating and cooling phase of the experiment. Neither the extrapolation number, n, nor its reciprocal varied linearly with the reciprocal absolute temperature (Fig. 5A). Since the extrapolation number does not reflect a rate, the application of a rate theory to this parameter is probably inappropriate. The common practice, however, is to analyze the inactivation rate, $1/D_0$, while ignoring the second survival-curve parameter, n (Fig. 5B). This approach is equivalent to approximating a 'shouldered' survival curve by a simple exponential curve. Although acceptable for specific analyses in specific circumstances, the approach will be misleading when the information is used to extrapolate absolute cell killing at other temperatures.

An alternate description of hyperthermia cell survival curves can be based

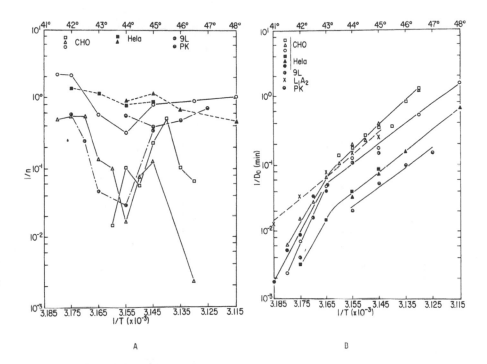

A   B

Figure 5: Arrhenius plot for the survival curve pararmeters, n and D_0. The graph was constructed from data in the literature after correction for the temperature transients during the heating and cooling phase. Survival points for each curve were fitted by a Chi-square minimization procedure to obtain newly calculated survival curve parameters. The calculated extrapolation number, n, (Panel A) and the D_0 (Panel B) were then used to construct the Arrhenius plot (Henle and Dethlefsen 1980; reprinted with permission).

on the linear-quadratic survival model (Chadwick and Leenhouts 1973). The principal disadvantage of this model is the relative difficulty of visualizing survival curves on the basis of specific values for alpha and beta. An Arrhenius analysis of alpha and beta, however, yielded a linear relationship with the reciprocal absolute temperature (Fig. 6; Roti Roti and Henle 1980), although the survival data used for constructing both Figs 5 and 6 were the same. In this case, the parameter alpha, but not beta, represents an exponential rate constant; beta is equivalent to a probability constant for a two-step process. Although the linear relationships between temperature and the two survival-curve parameters can be used to derive an adequate description of temperature-dependent cell survival, the meaning of the slope for beta in the Arrhenius formalism is obscure. However, its simple relationship to $(1/T)$ is suggestive of a useful relationship to temperature.

The application of an Arrhenius analysis for identifying rate-limiting steps in mechanisms of heat killing violates basic thermodynamic assumptions of reversibility (Henle 1983). The phenomenon of cell death is so irreversible that even the assumption of 'quasi-irreversibility' is clearly violated. Furthermore the units inherent in the activation enthalpy and entropy entail the

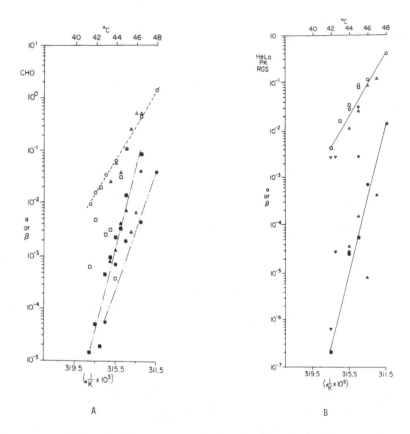

Figure 6: Arrhenius plot for the survival curve parameters, alpha and beta. Using the same published and corrected survival data as in Figure 5, the best fit was determined to the linear-quadratic cell survival model. Using the parameters alpha (open symbols) and beta (closed symbols), no inflection points can be observed. Panel A: data for CHO cells; Panel B: data for HeLa, pig kidney and rat gliosarcoma (9L) cells. Each set of symbols represent a specific source of data (Roti Roti and Henle 1980; reprinted with permission).

chemical concept of 'moles' which has no clear meaning when applied to living cells. In spite of these conceptual obstacles, the inflection point near 43 °C in Fig. 5B has been used to implicate different, albeit unknown, mechanisms of cellular heat death above and below 43 °C. The lack of inflection points near 43 °C for Arrhenius plots based on the linear-quadratic model suggests, however, that inflection points in Fig. 5B are peculiar to the specific representation of cell survival, i.e. the exclusive use of the terminal exponential slope of the cell survival curve. A valid thermodynamic analysis of cellular heat death is possible, but requires either calorimetric measurements on heated cells (Cheng *et al.* 1987; Lepock *et al.* 1987), or the application of rate theory to defined subcellular systems that satisfy at least a requirement of quasi-reversibility.

Without a thermodynamic justification, the Arrhenius plot of survival-curve parameters becomes a simple representational device of time–temperature relationships. A simple graphical representation of temperature-dependent cell killing, however, does not require an Arrhenius format. Instead, a linear or nonlinear plot of any specific survival endpoint, e.g., time/temperature to kill 90 per cent of cells, D_0, alpha, etc., versus temperature is preferable for ease of retrieving graphical data (Henle 1983).

2. The time course for hyperthermic cell death

As shown above, hyperthermia causes a dose-dependent fraction of cells to be killed, as measured by clonogenic assays of the type developed by Puck and Marcus (1956). These assays give no information regarding the events (and their timing) which accompany cell death and lysis. Obtaining such information requires different techniques. For example, cinematography, dye exclusion, and monolayer detachment have been used in studies of cell death following hyperthermia. The most complete study so far is that of Zielke-Temme and Hopwood (1982a, b) using cinematography to study the time course of cell death in CHO cells heated in G_1 phase. These studies showed that after a heat exposure which killed 94 per cent of the cells, the modal time for cell death measured in terms of rounding and cell lysis, was 28–32 hours (Fig. 7). Although this time interval is approximately the time required for two cell cycles, very few of the heated cells divide prior to cell death. Thus, this type of cell death does not appear to be associated with a mitotic

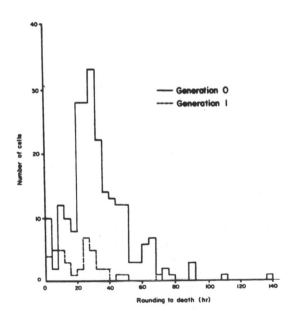

Figure 7: The time of death in relation to the previous mitosis. CHO cells were heated for 17.5 minutes at 45.5 °C. The time of death was recorded for cells that died following both regular and irregular divisions. The intervals are grouped into 4 hour segments (Zielke-Temme and Hopwood 1982; used with permission).

event. A similar observation is made when CHO cell death is monitored by detachment from the monolayer and compared with the mitotic delay (Leeper 1985). In contrast to the case for G_1 cells, further cinematography studies showed that most cells (>80 per cent) heated in S-phase completed the first mitotic event subsequent to heat (Coss and Dewey 1983). Thus, the time course of cell death following heat exposure is complex and our knowledge of it is still incomplete.

Cell death following ionizing radiation can be clearly divided into two types, reproductive death and interphase death (Altman 1970). At lethal radiation doses (20 to 40 Gy) cell killing occurs by reproductive death while at superlethal doses (>5000) cells die by interphase death. With high doses that kill 90 to 99 per cent of the cells, the following are characteristics of heat-induced cell death which are similar to those for radiation-induced reproductive death. Cell killing follows a sigmoidal survival curve and death occurs several hours to a few days after exposure. Unlike reproductive cell death, heat-induced cell death is not associated with a mitotic event. However, if one observes death of the cells that complete the first mitosis, then it appears that the death of these cells appears to be associated with a mitotic event. After mitosis these dying cells exhibit fusions and multipolar divisions (Zielke-Temme and Hopwood 1982) as well as lethal sectoring (Fig. 8, Rice et al. 1984). Therefore, heat-induced cell death is only partially similar to radiation-induced reproductive cell death at lethal doses. Alternatively, it is possible that G_1 cells die by an interphase death mechanism while S and G_2 cells die by a reproductive death mechanism. Also, it appears that heat causes interphase death at higher survival levels than does radiation (i.e. lethal as opposed to superlethal).

E. THE ROLE OF PROLIFERATIVE STATUS IN HEAT SENSITIVITY

1. Cell cycle-dependent thermal sensitivity

One of the principal motives that rekindled interest in hyperthermia as an adjuvant cancer treatment modality was the observation that the cell-cycle age response to heat was complimentary to that of radiation (Westra and (Dewey et al.1971; Westra and Dewey 1971; King et al. 1980; Read et al. 1984). However, HeLa mitotic cells appear to be most sensitive to both hyperthermia (Fox et al. 1985) and X-rays (Terasima and Tolmach 1963). Heat sensitivity through the remainder of the HeLa cell cycle shows a pattern similar to that for hamster cells (Figs 9 and 10; Read et al. 1984; Fox et al. 1985). The sensitivity varies by more than 10-fold across the cell cycle and most of this variation appears due to a modification in the extrapolation number on the heat survival curve (Dewey et al. 1971). However, it is important to note that due to the >10-fold difference in sensitivity between G_1 and S cells, a small G_1 contamination could dominate the final slope of the survival curve and contribute to the apparent heat resistance of S and G_2 populations. Because of this problem an extra effort has been made to obtain pure cell populations. These methods include fluorescent activated

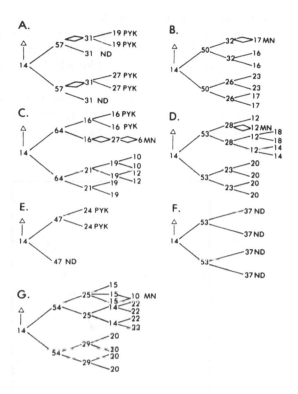

Figure 8: Representative CHO cell pedigrees as obtained by phase-contrast time-lapse cinematography. The numbers correspond to the time in hours to division; PYK, pyknotic; MN, multinucleate; ND nondividing. All others were judged to be viable (i.e., eventual division to a colony of 50 cells or larger). All cells were heated (45.5 °C for 20 minutes) in the first generation (from Rice *et al.* 1984; used with permission).

cell sorting of S-phase cells (Rice *et al.* 1986), ^{3}H-TdR suicide (Mackey and Dewey 1988) and the use of bivariate, DNA-BUdR analysis of cell progression (Rice *et al.* 1986; Mackey and Dewey 1987). These results indicate that the thermal sensitivity of S-phase cells is generally underestimated by common synchronization methods.

Since cells in G_1 phase are the most thermally resistant subpopulation, it is relevant to ask whether the pattern of cellular heat sensitivity across G_1 is related to the duration of the G_1 phase. As many readers will know, the pattern of G_1 radiosensitivity is a function of the length of the G_1 phase. Cells with a long G_1 (HeLa) show a peak of radioresistance in early G_1 (Terasima and Tolmach 1963); while cells with a short G_1 (CHO) show little or no variation in sensitivity across G_1 (Sinclair and Morton 1966). This question has not been directly addressed for hyperthermia. However, the data shown in Figs 9 and 10 suggest that cellular heat sensitivity is relatively constant across G_1 in CHO cells which have a relatively short G_1 and show a peak of thermal resistance in mid G_1 of HeLa cells which have a long G_1. Thus, the thermal resistance pattern in G_1 appears similar to that for radiation. However, variation of sensitivity across G_1 was observed for mouse neuro-blastoma cells (van Dongen and van Wijk 1986) which have a 3.5 hour G_1. phase which is shorter than that for CHO. Interestingly, this variation in sensitivity was dependent upon the presence of serum in the culture medium. Thus, verification of the role of G_1 duration in the pattern of thermal sensitivity and its implications for the mechanism for cell killing requires further data.

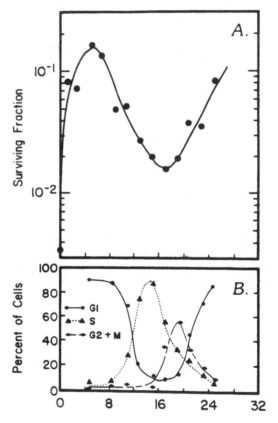

Figure 9: (A) Survival of HeLa cells following a 30-minute, 45.0 °C-heat 'dose' delivered at various times after release from mitosis. (B) Fractions of the cell populations in specific phases of the cell cycle at various times after mitosis were determined by computer analysis of FCM profiles (Fox *et al.* 1985; used with permission).

2. *Heat-induced division delays and redistribution*

Hyperthermia causes alterations in the normal proliferation patterns of cycling cells. Some of these alterations appear immediately during and after heating while others develop more slowly with time after heat. The predominant methods of study have been based on flow cytometry and include univariate

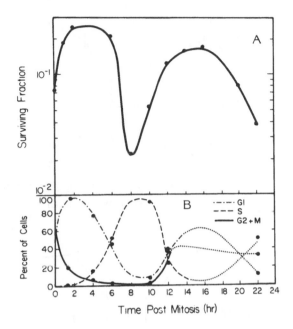

Figure 10: (A) Survival of CHO cells following a 20 minute heat dose at 45 °C at various times after mitosis. (B) Fractions of the cell population in various cell-cycle stages were determined by flow cytometry analysis. The dotted lines are 'plausible interpolations' [*sic*] of the data between 12 and 22 hours (Read *et al.* 1984; used with permission).

DNA analysis (Kal *et al.* 1975; Kal and Hahn 1976; Sapareto *et al.* 1978), bivariate DNA and antiBUdR analysis (Rice *et al.* 1984, 1986), and bivariate DNA and nuclear protein analysis (Roti Roti *et al.* 1986). Immediately following a heat exposure, cell-cycle progression is halted or greatly slowed uniformly throughout the cell cycle. The cell-cycle arrest is reflected by the near variance of cell-cycle phase fractions prior to the resumption of cell multiplication and cell loss due to interphase death (Fig. 11). This condition persists for a heat-dose dependent time interval. For example, in HeLa cells 90 per cent of the G_1 cells remain in G_1 for up to 7 hours following 45 °C, 30 minutes and for up to 4 hours following 43 °C, 60 minutes (Roti Roti *et al.* 1986). Cells in S-phase are the first to recover from the overall cell-cycle delay. The fraction of cells in G_2 begins to increase 2 to 4 hours before the fraction of cells in G_1 changes. At 6 to 8 hours (depending upon the specific heat exposure) following a heat shock, cells begin to progress out of G_1, proceed through S and accumulate in G_2. The fraction of cells in G_2 reaches a maximum between 10 to 15 hours after 43 °C, 60 minutes and between 20 to 30 hours after 45 °C, 30 minutes. Cell division appears to resume just after the G_2 maximum. However, the division rate was very low following recovery from the higher heat 'dose', presumably reflecting a fraction of cells that die prior to the completion of the first post-heat mitotic cycle. Thus, the effects of hyperthermia on cell-cycle progression can be summarized as an immediate arrest throughout the cell cycle followed by a G_2 block. The

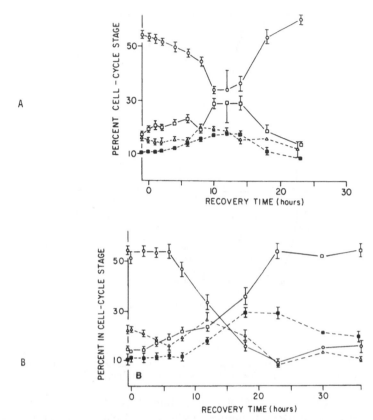

Figure 11: Cell cycle redistribution and nuclear protein content following (A) 45 °C for 60 minutes or following (B) 45 °C for 30 minutes. At various times (abscissa) following hyperthermia, the fraction of cells in each cell cycle stage was determined by DNA content analysis (measured flow cytometrically by PI fluorescence). The plotted points represent the average of eight pooled experiments (Roti Roti *et al.* 1986; used with permission), circles, G_1; triangles, early S; solid squares, late S; open squares, G_2.

heat-induced G_2 block has similarities with the X-ray-induced G_2 block. Both blocks are reversed by caffeine (Higashikubo *et al.* 1987) and both blocks are accompanied by changes in nuclear protein content (Roti Roti *et al.* 1986a, b).

The foregoing considerations include the cell population as a whole. The question arises whether post-hyperthermia proliferation kinetics differ for cells destined to survive and those destined to die. At this point there is no study which directly addresses this problem. However, one study compared the cellular multiplicity and division delay of the total population, as well as the viable and the nonviable subpopulations (Fig. 12; Rice *et al.* 1984) and those data give some insight into this problem. The division delay was found to be approximately equal for all three groups. Taken together with the G_2-block data, this result implies that both surviving and nonsurviving cells are equally delayed in G_2. However, direct experiments will be required to confirm this notion. After the delay, the nonviable cells showed a much slower rate of cell division than that for the viable and total cell populations (Fig. 12). This result could be due to a lengthening of the cell-cycle transit time and/or due to the loss of dying cells.

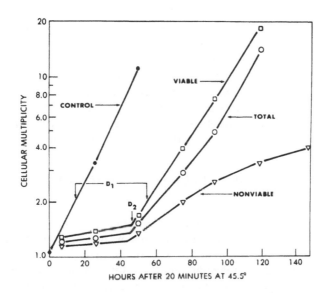

HOURS AFTER 20 MINUTES AT 45.5°

Figure 12: Division delay and subsequent growth kinetics of viable and nonviable CHO cells following 20 minutes at 45.5 °C. D_1 and D_2 refer to different methods of defining the length of division delay, i.e. the time for the heated population to reach the same multiplicity as an arbitrary control point (D_1), or the time during which there is no increase in multiplicity (D_2). Viability was defined as the ability of a cell to form a microcolony of 50 cells or more (Rice *et al.* 1984; used with permission).

3. Exponential growth versus plateau phase; P versus Q

When cells are present in tissues *in vivo*, as opposed to cells growing in culture, a large cell fraction is in a nonproliferating or quiescent state (Q). Proliferation status is known to affect the sensitivity of cells to various therapeutic agents (Valeriote and van Putten 1975). Therefore, it is important to consider potential differences in the thermal sensitivity between proliferating (P) and Q cells. The Q state has been characterized by several criteria listed below (Epifanova 1977; Gelfant 1977; and Dethlefsen 1980). Lack of a clear definition of the Q state and its relationship to growth arrest achieved by various environmental conditions appears to be at least one of the reasons for conflicting results in the literature regarding the thermal sensitivity of Q cells (reviewed in Hahn 1982). Certain cell lines derived from a mouse mammary tumor are known to form the Q state *in vitro* with ≥97 per cent of the cells with G_1 DNA content and ≤2.5 per cent of the cells incorporating ³H-TdR during a labeling period equivalent to two generation times (Wallen *et al.* 1984a, b). When P and Q cells were heated in their respective environmental conditions (i.e. relatively complete medium at pH 7.2 and depleted medium at pH 6.4 to 6.6), their heat sensitivities are similar (Wallen *et al.* 1984c). However, when the Q cells were placed in fresh medium, they were found to be approximately two-fold more heat resistant than P cells. This change in heat resistance developed with time after the medium change and appeared to be dependent upon the glucose concentration and pH of the culture medium (Wallen and Gutierrez 1987). Another relevant, but contrasting observation is that mouse SCK tumor cells *in situ* are more heat sensitive than the same cells grown in culture (Rhee *et al.* 1986). The transition from the *in vivo* thermal sensitivity to the higher heat resistance in culture required about 3 hours, implying that growth conditions were a critical factor in the differential sensitivity of these tumor cells. The results emphasize the importance of the cellular growth state in thermal sensitivity and the need for more information on the role of the cellular environment.

F. ALTERED STATES OF CELLULAR HEAT SENSITIVITY

1. Environmental factors
Cellular heat sensitivity is also dependent on the cells' environment in other ways. High concentrations of small molecular weight solutes in the medium can alter the aqueous environment of intracellular macromolecules and either augment or reduce macromolecular heat stability. Although this subject is reviewed in detail elsewhere within this volume, it is appropriate to mention the concept of solute-induced shifts in the 'biologically equivalent' temperature. For example, heat sensitization by compounds, such as ethanol, need not be interpreted in terms of a selective partitioning into cellular membranes or in terms of a specific chemical effect, such as oxidation to acetaldehyde. Instead, the addition of ethanol to the aqueous medium can alter the solvent properties of water in a way that affects the stability of all macromolecules and their interactions with each other. For example, ethanol plus heat was observed to increase the amount of heat-induced excess protein in isolated nuclei (Roti Roti and Wilson 1984). Thus, cells in ethanol at 37 °C can behave like cells in medium without ethanol that were maintained at a higher temperature (Henle *et al.* 1986). By inference, polyhydroxy compounds, such as polyols, have the opposite effect. The non-specific solvent effect on macromolecular heat stability has been demonstrated with model proteins (reviewed in Henle *et al.* 1986) and can have the dual effect of impairing protein function while increasing molecular stability. For example, ethanol at mildly hypothermic temperatures gives rise to heat sensitization that resembles stepdown heating, but without a prior heat shock (Henle *et al.* 1986) and may reflect an inhibition of the cellular capacity for repair of heat damage.

2. Cell density and attachment
Cellular heat sensitivity is modified not only by exogenous additives, but also the cellular interactions with each other and their substrate. Cell density and cellular attachment to substrate has only a limited effect on growth of

many cell lines under normal culture conditions, but can significantly affect cell survival during hyperthermia. Cellular exposure to trypsin, cell scraping closely before or after hyperthermia, or heating cells in suspension, rather than in monolayer can potentiate heat killing in a major way (Highfield *et al.* 1984; Warters and Henle 1982). In contrast, cell density generally has little effect on heat sensitivity; in CHO cells, an enhanced heat sensitivity was demonstrated only when cellular multiplicity was less than two (Highfield *et al.* 1984). A multiplicity of approximately two is a standard condition in survival experiments when CHO cells are plated 16 hours (overnight) prior to hyperthermia.

3. Thermotolerance and energy substrates

Thermotolerance is a major modifier of cellular heat sensitivity. An extensive review of this phenomenon has been published elsewhere (Henle 1987). Likewise, conditions that impair thermolerance development could be major modifiers of cellular heat sensitivity. Although inhibition of thermotolerance does not occur easily, an impaired cellular energy status is likely to sensitize non-tolerant cells (Nagle *et al.* 1982) and prevent development of tolerance in appropriately conditioned cells (Henle 1987). Hyperthermia *per se* does not cause cellular depletion of high-energy phosphates (Henle *et al.* 1984).

4. Stepdown heating

Cellular heat sensitivity can be increased under conditions of stepdown heating (Henle and Leeper 1976). Stepdown heating is accomplished by an initial acute high-temperature shock, followed by low-temperature hyperthermia. Prolonged exposure to low-temperature hyperthermia, by itself, is either non-lethal or relatively ineffective for heat killing (Henle and Leeper 1976; Joshi and Jung 1979; Miyakoshi *et al.* 1979; Henle 1980; Rofstad and Brustad 1986). The stepdown heating phenomenon is illustrated in Fig. 13 for a number of clones derived from a single human melanoma xenograft (Rofstad and Brustad 1986). Initial heat conditioning was accomplished by a heat treatment of 90 minutes at 43.5 °C, followed by chronic hyperthermia up to 4 hours at 41.5 °C.

The melanoma data illustrate both the degree of heterogeneity in cellular heat sensitivity and in the magnitude of heat sensitization by stepdown heating within clones from the same tumor. In spite of this heterogeneity, none of these clones failed to exhibit sensitization by stepdown heating entirely, suggesting that the basic phenomenon reflects a fundamental aspect of the cellular heat response. The authors calculated a sensitization ratio based on slopes derived from the survival curves analogous to the thermotolerance ratio; these ratios ranged from approximately 5 up to a value of 15, but did not correlate either with cellular heat sensitivity, i.e. survival after heat conditioning, or with the cellular capacity for thermotolerance development (Rofstad and Brustad 1986). The interaction between the high-temperature shock (90 minutes, 43.5 °C), and low-temperature hyperthermia (41.5 °C), decayed fairly rapidly both in CHO cells (Henle 1980) and in human melanoma cells (Rofstad and Brustad 1986) with a half time of 1 to 3 hours. These decay kinetics resemble those for the repair of sublethal hyperthermia damage (Henle 1980; Nielsen *et al.* 1982; Jung 1986).

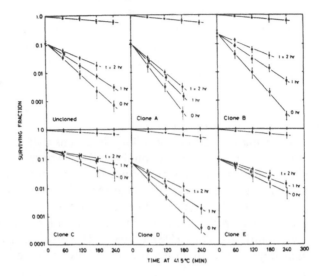

Figure 13: Sensitization to 41.5 °C by stepdown heating in various cell clones derived from a single human melanoma xenograft. The curve with solid circles at the top of each panel shows the survival response to 41.5 °C alone; the other survival curves in each panel were obtained from cells with heat conditioning of 90 minutes at 43.5 °C followed at 0, 1, or 2 hours later by graded exposures to 41.5 °C hyperthermia. Cell survival was measured by colony-forming ability in soft agar with control plating efficiencies of 10 to 50 per cent (Rofstad and Brustad, 1986; reprinted with permission).

The Arrhenius plot in Figure 14 can be viewed as a map for the temperature dependence of the stepdown heating effect (Henle 1980). The graph shows that the magnitude of sensitization to low-temperature hyperthermia increases with decreasing temperatures below 43 °C. At 42 °C, the ratio of slopes from survival curves with or without prior heat shock is relatively small (approximately 3), but approaches a value of 20 near 40 °C. Dashed lines, running approximately parallel to the stepdown heating line from 39 °C to 43 °C show that even with heat conditioning at or below 43 °C, e.g., 42 °C, cellular heat sensitivity is increased by stepdown heating. Heat conditioning above 43 °C at isotoxic levels caused the same extent of heat sensitization (Henle 1980). Cell survival after heat conditioning, i.e. the conditioning heat 'dose', had only a minor effect on the degree of heat sensitization, and then only when the conditioning heat 'dose' was very low (inset, Fig. 8).

Mechanisms of the stepdown heating phenomenon, like that of thermotolerance, remain unknown. Nevertheless, investigators have speculated on the potential relationship between heat sensitization via stepdown heating and the inhibition of thermotolerance development. However, the stepdown heating effect can be demonstrated in thermotolerant cells (Henle 1987) and even in cell lines e.g., L1A2, where thermotolerance development does not occur during chronic low-temperature hyperthermia (Nielsen *et al.* 1982). Conversely, thermotolerance development has been observed under stepdown heating conditions (Henle *et al.* 1978; Henle 1980). Alternate hypotheses for mechanisms of the stepdown heating effect are based on the concept of thermal damage repair (Henle and Leeper 1976; Jung 1986). The nature of the postulated sublethal heat damage, however, remains undefined. The potential clinical significance of stepdown heating is still unclear; a more detailed discussion of biological aspects of this phenomenon can be found elsewhere (Henle 1987).

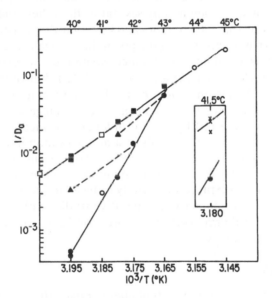

Figure 14: Arrhenius plot for stepdown heating with CHO cells. For stepdown heating, cells were heated for 10 minutes at 45 °C and then transferred immediately to the lower hyperthermic temperature (squares). Control cells that were not stepdown heated are denoted by circles. Triangles indicate results with stepdown heating for heat conditioning at 43 °C or 42 °C; the dashed lines connect the conditioning and the chronic, lower maintenance temperature. The inset shows the effect of heat conditioning with 5 minutes, 45 °C (lowest x) and 10 or 20 minutes, 45 °C heat conditioning (continuous x on the line) on the value for D_0 at 41.5 °C. Open or closed symbols refer to the source of the data (Henle 1980; reprinted with permission).

5. Mutants in cellular heat sensitivity

In principle, comparative studies with mutants in cellular heat sensitivity should identify specific aspects of mechanisms leading to cellular heat death and thermotolerance. However, such studies have only begun. The initial paper by Harris (1980) was followed by a number of other studies aimed at isolating mutants either with variable heat sensitivities and differing capacities for thermotolerance developments (Laszlo and Li 1985; Anderson *et al.* 1986; Landry and Chretien 1986;; Harvey and Bedford 1987).

Harris (1980) isolated clones of stable heat-resistant mutants of V79 Chinese hamster lung fibroblasts and characterized these in terms of growth rates and the morphology of colonies formed from plated cells. Heat-resistant mutants and wild-type sensitive cells were fused either as whole cells or as anuclear cytoplasts. Cytoplasts from sensitive cells did not alter the heat sensitivity of heat-resistant cells, suggesting that heat resistance was a nuclear characteristic. Fusions between sensitive and resistant whole cells yielded hybrids that were only slightly more heat-resistant than sensitive cells, suggesting that heat resistance was either recessive or weakly condominant (Harris 1980).

The expression of heat-shock proteins and/or the capacity for thermotolerance development was examined by several of the other studies (Laszlo and Li 1985; Anderson *et al.* 1986; Landry and Chretien 1986). These investigators isolated their own mutants, raising the possibility that specific findings applied only to each separate strain. For instance, some mutants expressed elevated levels of heat-shock proteins (Laszlo and Li 1985), but other heat-resistant mutants did not (Anderson *et al.* 1986). The interpretation of such variable results remains tentative, but has been used to suggest that cellular heat resistance can be conferred by several independent factors. Cholesterol-to-phospholipid ratios were altered only in a minor fashion in the heat-resistant mutants described by Anderson *et al.* (1986). To date, no mutants have been isolated that are incapable of developing thermotolerance, although the magnitude of thermotolerance was greatly reduced in specific mutant (Harvey and Bedford 1987). In summary, comparative studies with mutants of cellular heat sensitivity and thermotolerance expression have not yet made a major impact in thermal biology.

G.   SUMMARY

Cellular heat sensitivity is an inherent characteristic that cannot be predicted on the basis of measured sensitivity to other cytotoxic agents. Although hyperthermia has a profound effect on all aspects of cellular biochemistry and function, quasi-thermodynamic analyses of heat survival curves provide little information to distinguish between lethal and sublethal heat lesions. Cellular heat sensitivity is dependent on environmental conditions during hyperthermia, but cells can also respond to sublethal heat damage by altering their intrinsic sensitivity to hyperthermia.

ACKNOWLEDGEMENT

The authors would like to thank Ms Kathy McDonald for preparation of this manuscript and Dr Andrei Laszlo for critical comments. Dr Joseph L. Roti Roti was supported by NCI grant numbers CA 41102 and CA 43198. Dr Kurt J. Henle was supported by NCI grant numbers CA 33405 and CA 35689.

REFERENCES

Altman, K.I., Gerber, G.B., and Okada, S. (1970). Radiation-induced death. In *Radiation Biochemistry*, Vol. 1, Academic Press, New York, p. 247.

Anderson, R.L., Tao, T.W., Betten, D.A. and Hahn, G.M. (1986). Heat shock protein levels are not elevated in heat-resistant B16 melanoma cells. *Radiat. Res.* **105**: 240–246.

Blackburn, M.J., Wheldon, T.E., Field, S.B., and Goldman, J.M. (1984). The sensitivity to hyperthermia of human granulocyte/macrophage progenitor cells (CFU-GM) derived from blood or marrow of normal subjects and patients with chronic granulocytic leukemia. *Br. J. Cancer* **50**, 745–751.

Borelli, M.J., Wong, R.S.L., and Dewey, W.C. (1986). A direct correlation between hyperthermia-inducing membrane blebbing and survival in synchronous G_1 CHO cells. *J. Cell Physiol.* **126**, 181–190.

Bowler, K., Duncan, C.J., Gladwell, R.T., and Davison, T.F. (1973). Cellular heat injury. *Comp. Biochem. Physiol.* **45A**, 441–450.

Bromer, R.H., Mitchell, J.B., and Soares, N. (1982). Response of human hematopoietic precursor cells (CFUc) to hyperthermia and radiation. *Cancer Res.* **42**, 1261–1265.

Cavaliere, R., Ciocatto, E.C., Giovanella, B.C., Heidelberger, C., Johnson, R.O., Margottini, M., Mondovi, B., Moricca, G., and Rossi-Fanelli, A. (1967). Selective heat sensitivity of cancer cells. Biochemical and clinical studies. *Cancer* **20**, 1351–1381.

Chadwick, K.H., Leenhouts, H.P. (1973). A molecular theory of cell survival. *Phys. Med. Biol.* **18**, 78–87.

Cheng, K-H, Hui, S.W., Lepock, J.R. (1987). Protection of intracellular Ca-channel and Ca-ATPase from hyperthermia. *Book of Abstracts, 35th Annual Meeting, Radiation Research Society/North American Hyperthermia Group*, Atlanta, GA, Feb. 21–26, #Be–6, pg. 22.

Coss, R.A., and Dewey, W.C. (1983). Mechanism of heat sensitization of G_1 and S phase cells by procaine-HCl. In *Tumour Biology and Therapy, Proceedings of the 7th International Congress of Radiation Research*, J.J. Broerse, G.W. Barendsen, H.B. Kal, and A.J. van der Kogel, (Eds.), Martinus Nijhoff, Dordrecht.

Dardalhon, M., More, C., Averbeck, D., and Berteaud, A.J. (1983). Microwave-induced hyperthermic effects on the cytoplasm of Chinese hamster V79 cells detected by fluorescence polarization measurements. In *Tumour Biology and Therapy*, pp. D6-06-D6-07. Broerse, J.J., Barendsen, G.W., Kal, H.B. and Van Der Kogel, A.J. (Eds.), Martinus Nijhoff, Dordrecht.

Dethlefsen, L.A. (1980). In quest of the quaint quiescent cell. In *Radiation Biology in Cancer Research*, p. 415. Meyn, R.E. and Withers, H.R. (Eds.), Raven Press, New York.

Dewey, W.C., Westra, A., and Miller, H.H. (1971). Heat-induced lethality and chromosomal damage in synchronized Chinese hamster cells treated with 5-bromodeoxyuridine. *Int. J. Radiat. Biol.* **20**, 505–520.

Elkind, M.M., and Whitmore, G.F. (1967). *The Radiobiology of Cultured Mammalian Cells*, Gordon and Breach Science Publishers, New York, pp. 16–17.

Epifanova, O.I. (1977). Mechanisms underlying the differential sensitivity of proliferating and resting cells to external factors. *Int. Rev. Cytol. (suppl. 5)*, 303.

Fajardo, L.F., Schreiber, A.B., Kelley, N.I., and Hahn, G.M. (1985). Thermal sensitivity of endothelial cells. *Radiat. Res.* **103**, 276–285.

Fajardo, L.F., Kowalski, J., Prionas, S., Allison, A., and Kwan, H. (1987). Dose-dependent inhibition of angiogenesis by hyperthermia. (Submitted).

Fox, M.H., Read, R.A., and Bedford, J.S. (1985). The cell cycle dependence of thermotolerance. III. HeLa cells heated at 45 °C. *Radiat. Res.* **104**, 429–442.

Gelfant, S. (1977). Cycling-noncycling cell transition in tissue aging, immunological surveillance, transformation and tumor growth. *Int. Rev. Cytol.* **70**, 1.

Gerweck, L.E., and Burlett, P. (1978). The lack of correlation between heat and radiation sensitivity in mammalian cells. *Int. J. Radiat. Oncol. Biol. Phys.* **4**, 283–285.

Giovanella, B.C., and Mondovi, B. (1977). Introduction. In *Selective Heat Sensitivity of Cancer Cells*. Rossi-Fanelli, A., Cavaliere, R., Mondavi, B., and Moricca, G. (Eds.), Recent Results in Cancer Research, Vol. 59, pp. 1–6, 20–22.

Hahn, G.M. (1982). *Hyperthermia and Cancer*, Plenum Press, New York, p. 23.

Harris, M. (1980). Stable heat-resistant variants in populations of Chinese hamster cells. *J. Natl. Cancer Inst.* **64**, 1495–1501.

Harvey, W.F., and Bedford J.S. (1987). Isolation and preliminary characterization of thermotolerant defective, heat sensitive mutants of CHO cells. *Book of Abstracts, 35th Annual Meeting*

Radiation Research Society/North American Hyperthermia Group, Atlanta, GA, Feb. 21–26, #Ac–17, p. 11.

Heilbrunn, L.V. (1954). Heat death. *Scientific American*, **190**, 70–74.

Henle, K.J. (1980). Sensitization to hyperthermia below 43 °C induced in Chinese hamster ovary cells by step-down heating. *J. Natl. Cancer Inst.* **64**, 1479–1483.

Henle, K.J. (1982). Thermotolerance in the murine jejunum. *J. Natl. Cancer Inst.* **68**, 1033–1036.

Henle, K.J. (1983). Arrhenius analysis of thermal response. In *Hyperthermia in Cancer Therapy*, Storm, F.K. (Ed.), G.K. Hall Medical Publishers, Boston, pp. 47–53.

Henle, K.J. (1987). Thermotolerance in cultured mammalian cells. In *Thermotolerance*, Vol. I, pp. 13–71. Henle, K.J. (Ed.), CRC Press, Boca Raton, FL.

Henle, K.J., and Dethlefsen, L.A. (1980). Time-temperature relationships for heat-induced killing of mammalian cells. *Ann. New York Acad. Sci.* **335**, 234–253.

Henle, K.J., Karamuz, J.E., and Leeper, D.B. (1978). Induction of thermotolerance in Chinese hamster ovary cells by high (45 °) and low (40 °) hyperthermia. *Cancer Res.* **38**, 570–574.

Henle, K.J., and Leeper, D.B. (1976). Combinations of hyperthermia (40 °, 45 °C) with radiation. *Radiology* **121**, 451–454.

Henle, K.J., Moss, A.J., and Nagle, W.A. (1986). Temperature-dependent induction of thermo-tolerance by ethanol. *Radiat. Res.* **108**, 327–335.

Henle, K.J., Nagle, W.A., Moss, A.J., and Herman, T.S. (1984). Cellular ATP content of heated Chinese hamster ovary cells. *Radiat. Res.* **97**, 630–63r.

Higashikubo, R., Holland, J.M., and Roti Roti, J.L. (1987). Kinetic role of caffeine following radiation and hyperthermia. *Book of Abstracts, Cell Kinetics Society Annual Meeting*, Abst. 11, p. 46.

Highfield, D.P., Holahan, E.V., Holahan, P.K., and Dewey, W.C. (1984). Hyperthermic survival of Chinese hamster ovary cells as a function of cellular population density at the time of plating. *Radiat. Res.* **97**, 139–153.

Hume, S.P. (1985). Experimental studies of normal tissue response to hyperthermia given alone or combined with radiation. In *Hyperthermic Oncology 1984*, Vol. II, pp. 53–70. Overgaard, J. (Ed.), Francis & Taylor, London.

Joshi, D.S., Deys, B.F., Kipp, J.B.A., Barendsen, G.W., and Kralendonk, J. (1977). Comparison of three mammalian cells-lines with respect to their sensitivities to hyperthermia, gamma-rays and U.V.-radiation. *Int. J. Radiat. Biol.* **31**, 485–492.

Joshi, D.S., and Jung, H. (1979). Thermotolerance and sensitization induced in CHO cells by fractionated hyperthermic treatment at 38 °–45 °C. *Eur. J. Cancer* **15**, 345–350.

Jung, H., (1986). A generalized concept for cell killing by heat. *Radiat. Res.* **106**, 46–72.

Kal, H.B., Hatfield, M., and Hahn, G.M. (1975). Cell cycle progression of murine sarcoma cells after X-irradiation or heat shock. *Radiology* **117**, 215–217.

Kal, H.B., and Hahn, G.M. (1976). Kinetic responses of murine sarcoma cells to radiation and hyperthermia *in vivo* and *in vitro*. *Cancer Res.* **36**, 1923–1929.

Kalland, T., and Dahlquist, I. (1983). Effects of *in vitro* hyperthermia on human natural killer cells. *Cancer Res.* **43**, 1842–1846.

King, G.A., Archambeau, J.O., and Klevecz, R.R. (1980). Survival and phase response following ionizing radiation and hyperthermia in synchronous V79 and EMT6 cells. *Radiat. Res.* **84**, 290–300.

Landry, J., and Chretien, P. (1986). Chinese hamster heat-resistant mutant cells with altered growth properties. *Book of Abstracts, 34th Annual Meeting Radiation Research Society/North American Hyperthermia Group*, Las Vegas, NV, April 12–17, #Co–15, p. 49.

Laszlo, A., and Li, G.C. (1985). Heat-resistant variants of Chinese hamster fibroblasts altered in expression of heat shock protein. *Proc. Natl. Acad. Sci. USA*. **82**, 8029–8033.

Leeper, D.B. (1985). Molecular and cellular mechanisms of hyperthermia alone or combined with other modalities. In *Hyperthermic Oncology*, pp. 9–40. Overgaard, J. (Ed.), Taylor and Francis, London.

Lepock, J.R., Inniss, W.E., and Frey, H.E. (1987). DSC studies of the thermal denaturation of *Bacillus stearothermophilus* and *Bacillus megaterium:* relationship to maximum growth temperature. *Book of Abstracts, 35th Annual Meeting Radiation Research Society/North American Hyperthermia Group*, Atlanta, GA, Feb. 21–26, #Be–10, pg. 23.

Mackey, M.A., and Dewey, W.C. (1987). Cell cycle progression during chronic 42 °C hyperther-mia in S-phase CHO cells. *Book of Abstracts, 35th Annual Meeting Radiation Research Society/North American Hyperthermia Group*, Atlanta, GA, Feb. 21–26, p. 23.

Mackey, M.A., and Dewey, W.C. (1988). Time–temperature analyses of cell killing of synchron-ous G_1 and S phase Chinese hamster cells *in vitro*. *Radiat. Res.* **113**, 318–333.

Miyakoshi, J., Ikebuchi, M., Furukawa, M., Yamagata, K., Sugahara, T., and Kano, E. (1979). Combined effects of X-irradiation and hyperthermia (42 ° and 44 °C) on Chinese hamster V-79 cells *in vitro. Radiat. Res.* **79**, 77–88.

Mondovi, B., Strom, R., Rotilio, G., Agro, A.F., Cavaliere, R., and Fanelli, A.R. (1969). The biochemical mechanisms of selective heat sensitivity of cancer cells. I. Studies on cellular respiration. *Eur. J. Cancer* **5**, 129–136.

Nagle, W.A. (1987). The potential role of DNA damage and altered DNA metabolism in heat-induced cell killing and in cellular thermotolerance development. In *Thermotolerance*, Vol. II, pp. 19–46. Henle, K.J. (Ed.), CRC Press, Roca Baton, FL.

Nagle, W.A., Moss, A.J., and Baker, M.L. (1982). Increased lethality from 42 °C hyperthermia for hypoxic Chinese hamster cells heated under conditions of energy deprivation. *Natl. Cancer Inst. Monogr.* **61**, 107–110.

Nielsen, O.S., Henle, K.J., and Overgaard, J. (1982). Arrhenius analysis of survival curves from thermotolerant and step-down heated L1A2 cells *in vitro. Radiat. Res.* **91**, 468–482.

Puck, T.T., and Markus, P.I. (1956). Action of X-rays on mammalian cells. *J. Exp. Med.* **103**, 653–666.

Raaphorst, G.P. and Azzam, E.I. (1982). The thermal sensitivity of normal and ataxia telangiectasia human fibroblasts. *Int. J. Radiat. Oncol. Biol. Phys.* **8**, 1947–1950.

Raaphorst, G.P., Azzam, E.I., Borsa, J., Einspenner, M., and Vadaz, J.A. (1984). Inhibition of DMSO-induced differentiation by hyperthermia in a murine erythroleukemia cell system. *Can. J. Biochem. Cell Biol.* **62**, 1091–1096.

Raaphorst, G.P., Romano, S.L., Mitchell, J.B., Bedford, J.S., and Dewey, W.C. (1979). Intrinsic differences in heat and/or X-ray sensitivity of seven mammalian cell lines cultured and treated under identical conditions. *Cancer Res.* **39**, 396–401.

Raaphorst, G.P., Vadaz, J.A., and Azzam, E.I. (1984). Thermal sensitivity and radiosensitization in V79 cells after BrdUrd or IdUrd incorporation. *Radiat. Res.* **98**, 167–175.

Read, R.A., Fox, M.H., and Bedford, J.S. (1983). The cell cycle dependence of thermotolerance. I. CHO cells heated at 42 °C. *Radiat. Res.* **93**, 93–106.

Read, R.A., Fox, M.H., and Bedford, J.S. (1984). The cell cycle dependence of thermotolerance. II. CHO cells heated at 45 °C. *Radiat. Res.* **98**, 491–505.

Rhee, J.G., and Song, C.W. (1984). Thermosensitivity of bovine aortic endothelial cells in culture: *in vitro* clonogenicity study. In *Hyperthermic Oncology 1984*, Overgaard, J. (Ed.), Taylor and Francis, London.

Rhee, J.G., Song, C.W., Lyons, J.C., and Levitt, S.H. (1986). The intratumor environment prior to heating is a determinant of the solid tumor heat response. *Book of Abstracts, 34th Annual Meeting Radiation Research Society*, p. 19.

Rice, G.C., Gray, J.W., and Dewey, W.C. (1984). Cycle progression and division of viable and nonviable Chinese hamster ovary cells following acute hyperthermia and their relationship to thermal tolerance decay. *Cancer Res.* **44**, 1802–1808.

Rice, G., Laszlo, A., Li, G., Gray, J., and Dewey, W. (1986). Heat shock proteins within the mammalian cell cycle: relationship to thermal sensitivity, thermal tolerance, and cell cycle progression. *J. Cell. Physiol.* **126**, 291–297.

Rofstad, E.K., and Brustad, T. (1984). Response to heat treatment (42.5 °C) *in vivo* and *in vitro* of five human melanoma xenografts. *Br. J. Radiol.* **57**, 1023–1032.

Rofstad, E.K., and Brustad, T. (1986). Differences in thermosensitization among cloned cell lines isolated from a single human melanoma xenograft. *Radiat. Res.* **106**, 147–155.

Roti Roti, J.L., and Henle, K.J. (1980). Comparison of two mathematical models for describing heat-induced cell killing. *Radiat. Res.* **81**, 374–383.

Roti Roti, J.L., and Wilson, C.F. (1984). The effects of alcohols, procaine and hyperthermia on the protein content of nuclei and chromatin. *Int. J. Radiat. Biol.* **46**, 25–33.

Roti Roti, J.L., Uygur, N., and Higashikubo, R. (1986a). Nuclear protein following heat shock: protein removal kinetics and cell cycle rearrangements. *Radiat. Res.* **107**, 250–261.

Roti Roti, J.L., Kristy, M.S., and Higashikubo, R. (1986b). Nuclear protein content and cell progression kinetics following X-irradition. *Radiat. Res.* **108**, 52–61.

Roti Roti, J.L., and Laszlo, A. (1988). The effects of hyperthermia on cellular macromolecules. In *Hyperthermia and Oncology*, pp. 13–56. Urano, M. and Douple, E.B. (Eds), VSP, The Netherlands.

Reuifrock, A.C.C., Kanon, B., and Konings, A.W.T. (1987). Heat-induced K^+ loss, trypan blue uptake, and cell lysis in different cell lines: effect of serum. *Radiat. Res.* **109**, 303–309.

Sapareto, S.A., Hopwood, L.E., Dewey, W.C., Raju, M.R., and Gray, J.W. (1978). Effects of hyperthermia on survival and progression of Chinese hamster ovary cells. *Cancer Res.* **38**, 393–400.

Sinclair, W.K., and Morton, R.A. (1966). X-ray sensitivity during the cell generation cycle of cultured Chinese hamster cells. *Radiat. Res.* **29**, 450–474.

Strom, R., Santoro, A.S., Crifo, C., Bozzi, A., Mondovi, B., and Fanelli, A.R. (1973). The biochemical mechanism of selective heat sensitivity of cancer cells. IV. Inhibition of RNA synthesis. *Eur. J. Cancer* **9**.

Terasima, R., and Tolmach, L.J. (1963). X-ray sensitivity and DNA synthesis in synchronous populations of HeLa cells. *Science* **140**, 490–492.

Valeriote, F., and van Putten, L. (1975). Proliferation-dependent cytotoxicity of anticancer agents: A review. *Cancer Res.* **35**, 2619.

Van Dongen, G., and Van Wijk, R. (1986) Effect of serum on heat response of synchronized mouse neuroblastoma cells: protection of cell cycle progression, protein synthesis and survival. *Int. J. Radiat. Biol.* **50**, 77–91.

Van Zant, G., Flentje, D., and Flentje, M. (1983). The effect of hyperthermia on hemopoietic progenitor cells of the mouse. *Radiat. Res.* **95**, 142–149.

Wallen, C.A., Higashikubo, R., and Dethlefsen, L.A. (1984). Murine mammary tumour cells *in vitro*. I. The development of a quiescent state. *Cell Tissue Kinet.* **17**, 65–77.

Wallen, C.A., Higashikubo, R., and Dethlefsen, L.A. (1984). Murine mammary tumor cells *in vitro*. II. Recruitment of quiescent cells. *Cell Tissue Kinet.* **17**, 79–89.

Wallen, C.A., and Sullivan, M.D. (1984). Cytotoxicity of 45 °C hyperthermia in proliferating and quiescent mouse mammary carcinoma cells. *Book of Abstracts, 32nd Annual Radiation Research Society Meeting*, p. 49.

Wallen, C.A., and Gutierrez, R.H. (1987). Influence of physiological environment on the expression of thermotolerance in proliferating (P) and quiescent (Q) tumor cells. *Book of Abstracts, 35th Annual Radiation Research Society Meeting*, p. 11.

Warters, R.L., and Henle, K.J. (1982). DNA degradation in Chinese hamster ovary cells after exposure to hyperthermia. *Cancer Res.* **42**, 4427–4432.

Westra, A., and Dewey, W.C. (1971). Variation in sensitivity to heat shock during the cell cycle of Chinese hamster cells *in vitro*. *Int. J. Radiat. Biol.* **19**, 467–477.

Zielke-Temme, B., and Hopwood, L. (1982a). Time-lapse cinemicrographic observations of heated G1-phase Chinese hamster ovary cells. I. Division probabilities and generation times. *Radiat. Res.* **92**, 320–331.

Zielke-Temme, B., and Hopwood, L. (1982b). Time-lapse cinemicrographic observations of heated G1-phase Chinese hamster ovary cells. II. Cell death, fusion and multipolar divisions. *Radiat. Res.* **92**, 332–342.

Hyperthermia and Oncology, Vol. 1, pp. 83–98 (1988)
Urano and Douple (Eds)

Chapter 4.1

Modifiers of thermal effects: environmental factors

LEO E. GERWECK
Department of Radiation Medicine, Edwin L. Steele Laboratory, Massachusetts General Hospital, Harvard Medical School, Boston, MA 02114, USA

A. INTRODUCTION

There are currently two major rationale for hyperthermia in cancer therapy: differential heating and differential thermal sensitivity between tumor and normal tissue (Gerweck 1985). These rationale are based on differences in glycolytic and blood flow rates between tumor and normal tissue. Blood flow influences tissue response to heat in two respects; by dissipating energy from the heated volume, and by supplying and clearing cellular metabolites from tissue. Several of the metabolites supplied (oxygen and glucose) and cleared from tissue (CO_2, lactic acid) are known to affect the toxic response of cells to hyperthermia under *in vitro* conditions. However, tissue metabolite supply and clearance is not the sole regulator of tissue metabolic status. Blood flow rates in normal tissues vary enormously, i.e. by a factor of 100 or more, presumably in concert with the function and metabolic activity of the particular tissue. The metabolic status of tumor tissue is governed not only by blood flow characteristics of the tissue of origin, but also by any change in the metabolic characteristics of the tissue associated with malignant transformation.

This chapter will focus on the influence of nutrient and metabolite conditions on the response of cells to elevated temperatures, i.e. 41–46 °C. The first section will briefly review metabolic (glycolytic) differences between tumor and normal tissue (along with differences in tissue perfusion) which give rise to differences in the concentration of primary (e.g., H^+ and glucose) and secondary (ATP) metabolites which are modifiers of hyperthermic sensitivity. This will be followed by a discussion of each metabolite, i.e. its concentration in tumor vs normal tissue, and its impact on thermal sensitivity as revealed by *in vitro* studies.

B. GLYCOLYSIS IN TUMOR AND NORMAL TISSUE

Interest in the glycolytic characteristics of tumor and normal tissue can be traced to the pioneering work of Warburg (1926) who showed that tumor slices in general exhibited a higher rate of aerobic and anaerobic glycolysis than normal tissues. The studies of Warburg, contemporaries, and later workers have been exhaustively collected and analyzed by Aisenberg (1961), and reanalyzed by Weinhouse (1976). Animal and human tumors almost without exception display a high rate of acid production under hypoxic conditions (Table 1). Acid production, expressed as microliters of CO_2 given off per milligram dry weight per hour (Q), consistently exceeds 20 under anaerobic conditions, and the glycolytic rate of the human tumor cell is in the same range as that of animal tumors. Tumor slices also exhibit a considerable, but less consistent rate of aerobic glycolysis. Despite this variability, human and animal tumor aerobic glycolysis commonly exceeds a value of 10.

Table 1.
Condensed tumor metabolism data[a]

Tissue	N_2 Q_{CO_2}	O_2 Q_{CO_2}	Ref.
Fourteen normal tissues	7.2	2.1	Burk (1939)
of various animals	(2–19)	(0–10)	
Fifteen different tumor types	25.6	14.0	Burk (1939)
of various animals	(14.0–34.8)	(4.7–24.6)	
Thirteen human tumors	20.5	13.3	Warburg (1926)
	(13–29)	(5–19)	

Values are given as μl gas absorbed or evolved per mg dry weight with ranges given in parentheses.
[a] From Weinhouse 1976.

A limited number of normal tissues (mucous membrane of jejunum, kidney medulla, myeloid cells, and retina) exhibit high rates of aerobic and anaerobic glycolysis. Another group of tissues, including spleen, ovary, and endocrine organs, exhibit moderate levels of anaerobic glycolysis, but essentially no aerobic glycolysis. A large group of normal tissues e.g., liver, kidney, pancreas, thyroid gland, and submaxillary gland, undergo extremely limited aerobic glycolysis and little anaerobic glycolysis with Q values less than 5. Fewer studies of glycolytic rates have been made with homologous tumor and normal tissue. In contrast to normal liver, significant aerobic and anaerobic glycolysis is observed in rat hepatoma. However, in a series of well-differentiated mouse hepatomas, significant aerobic glycolysis was not observed. In studies with human skin and endometrium, the malignant tissue exhibited a high rate of glycolysis compared to homologous normal tissue. Leukemic lymph nodes generally exhibit an increased glycolysis compared to normal nodes. The leukocyte is one of the few normal tissues that exhibits a high rate of aerobic and anaerobic glycolysis. However, the aerobic and anaerobic glycolysis of leucocytes in chronic lymphocytic and myelocytic leukemia is depressed in comparison to the high value observed in normal white blood cells.

The bulk of the measurements dealing with the glycolytic activity of tumor and normal tissues were performed prior to the sixties, utilizing the Warburg mannometer in which the acid produced was measured as CO_2 released from a bicarbonate buffer. Generally speaking, acid production correlates with lactic acid production, however, a build-up of other acidic metabolites cannot be excluded in all cases. Finally, it should be pointed out that the studies reported here refer to tissue slices (primarily from rodent and human tissues) and should not be extrapolated to cultured cell lines examined under *in vitro* conditions. The etiology of the enhanced glycolysis of tumors is not known. A possibility is that the increase in glycolysis is an adaptive or auxillary response to an enhanced energy demand which occurs when slow or nonproliferative tissue is transformed to actively proliferating tissue.

In summary, the high rates of aerobic and anaerobic glycolysis is an impressive, although not a unique characteristic of tumor slices. Among the small group of normal tissues displaying a high aerobic and anaerobic glycolysis, bone marrow and tonsils may derive their glycolysis from constitutive white blood cells. Of the remaining normal tissues, only kidney medulla, retina, and the jejunal section of the intestinal mucosa, have been shown to exhibit high aerobic glycolysis. With the exception of a series of a few well-differentiated mouse tumors, the Q values for aerobic and anaerobic glycolysis usually exceed 10 and 20, respectively.

C. pH IN TUMORS

A possible consequence of an increased glycolytic rate of tumors is a reduction in tumor pH. The comprehensive compilation of human tumor and normal tissue pH measurements by Wike-Hooley *et al.* (1984) is shown in Fig. 1. This figure clearly shows that: (1) tumor pH is lower than normal subcutis or muscle pH (by an average of $\simeq$ 0.5 pH units); and (2) the range of pH values recorded is much greater in tumor than in normal tissue. These measurements were obtained with electrodes ranging in size from several micrometers to a few millimeters, and are generally considered to be representative of extracellular pH. In addition to the data shown in Fig. 1 Thistlewaite *et al.* (1985) has made 27 measurements of 14 human tumors of various histology with a modified 21 gauge needle electrode. Patients were those with primary recurrent or metastatic tumors which had failed previous therapy. The mean 'electrode pH' of these tumors of approximately 6.9 was also approximately 0.5 pH units below muscle and subcutis pH. While it is abundantly clear that tumor pH is lower than the pH of muscle and subcutis, the mean tissue pH values reported vary from investigator to investigator; for example, the mean pH values recorded by Wike-Hooley *et al.* (1984) in 32 various human tumors was 7.17, whereas Thistlewaite *et al.* observed a mean pH of approximately 6.9 in tumors of various histology. Because of

L. E. Gerweck

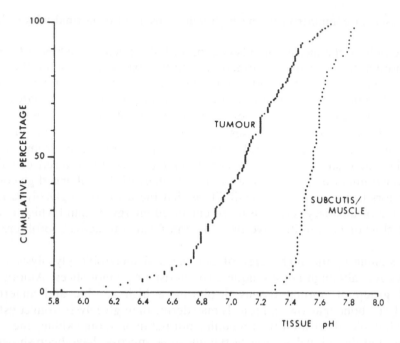

Figure 1: Cumulative distribution of human tumor pH determinations and corresponding values in normal subcutaneous or muscle tissue (Wike-Hooley *et al.* 1984).

these investigator-to-investigator differences, and the limited number of pH determinations made by the same investigator on various tumor types, it is unclear if tumor pH varies with tumor type. It is clear, however, that pH varies within the same tumor. An example is shown in Fig. 2 (Vaupel *et al.* 1981). These multiple point determinations were made with a 20 μm tip electrode in the C3H mammary carcinoma. Assuming these intratumor pH variations are maintained during hyperthermia, all areas of a particular tumor would not be equally susceptible to pH-modified heat damage.

In summary, the data presented indicate that the mean pH of human tumors is substantially lower than normal tissue pH. The mean pH of various human tumors is approximately 7.0, and may typically range from 6.5 to 7.4, within a particular tumor.

D. IMPACT OF pH ON CELL THERMAL SENSITIVITY

There are few *in vivo* studies which indicate whether or not naturally occurring low tumor pH enhances tumor response to hyperthermia. The difficulty in making this assessment is that it is not known how the same tumor would have responded to treatment if the pH conditions were other than naturally existed at the time of treatment. Another approach toward assessing the effect of pH on hyperthermic sensitivity under *in vivo* conditions involves a comparison of the thermal sensitivity of transplantable rodent tumors which have been acidified, e.g., via glucose injection, with those not receiving this

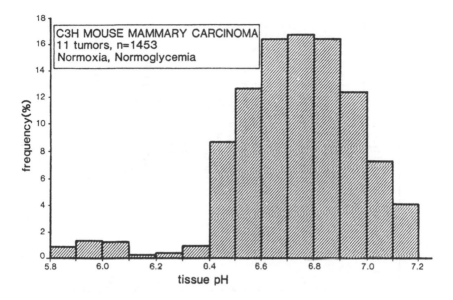

Figure 2: Frequency distribution of measured pH values by spear type microelectrodes (Vaupel *et al.* 1981).

pH-modifying treatment. Studies of this nature were first carried out by Von Ardenne (1972). Rats bearing the Ma 21224 mammary carcinoma were infused with glucose prior to and during a mild 41 °C, 70-minute hyperthermia. This treatment resulted in a 4 to 7 day growth delay in 63 per cent of the tumors. More recently, an enhancement of tumor sensitivity to hyperthermia by glucose injection or infusion has been demonstrated in a mouse fibrosarcoma by Urano *et al.* (1983), in a neural rat tumor by Jahde and Rajewsky (1983), and in a rat sarcoma and a carcinoma by Dickson and Calderwood (1983). Although it appears most likely that glucose increases tumor sensitivity by decreasing tumor pH, glucose is also known to substantially inhibit tumor blood flow, (Calderwood and Dickson 1983). A substantial inhibition of tumor blood flow may lead to the development of tissue hypoxia and glucose depletion during hyperthermia. The development of substantial nutrient restriction during hyperthermia may lead to thermal sensitization via cellular energy depletion as discussed later.

The most straight-forward demonstration of pH sensitization to hyperthermia arises from *in vitro* clonogenic assay studies. Advantages of this system arise from the fact that the pH of total cell population can be uniformly modified. This can be of critical assistance in data interpretation. Secondly, pH can be selectively altered without altering the concentration of other modifying metabolites. Finally, damage can be readily evaluated in terms of the loss of reproductive viability and avoids complications, such as cell division delay, etc. The effect of pH on cellular thermal sensitivity has now been examined in at least 10 different cell lines (see Table 2). Increased thermal

Table 2.
The influence of pH reduction upon the effectiveness of hyperthermia[a]

pH[b] effect	Para-[c] meters	Cell Line	Derivation	O_2[d] Status	Serum Present	Investigators
+	pHe	L1A2 ascites	Mouse lung	E	−	Overgaard (1976)
0	pHi	SDB	Rat mammary carcinoma	E	−	Dickson and Oswald (1976)
+	pHe	CHO	Chinese hamster	E	+	Gerweck (1977)
+	pHe	CHO	Chinese hamster	AH	+	Gerweck et al. (1979)
+	pHe	CHO	Chinese hamster	CH	+	Gerweck et al. (1979)
+	pHe	CHO	Chinese hamster	E	+	Gerweck et al. (1980, 1983)
+	pHe	Glioblastoma	Human	E	+	Gerweck and Richards (1981)
+	pHe	PNJ ascites	Mouse mammary carcinoma	E	+	Bichel and Overgaard (1977)
+	pHe	CHO	Chinese hamster	E	+	Freeman et al. (1977)
+	pHe	CHO	Chinese hamster	E	+	Freeman et al. (1980)
+	pHe	CHO	Chinese hamster	E	+	Freeman et al. (1981)
+	pHe	Madcap 37	Mouse mammary carcinoma	E	+	Meyer et al. (1979)
+	pHe	M 8013S	Mouse mammary carcinoma	E	−	Haveman (1979)
+	pHi	M 8013S	Mouse mammary carcinoma	EH	−	Haveman (1979)
0	pHe	CHO HAI	Chinese hamster	E	−	Li et al. (1980)
+	pHe	CHO HAI	Chinese hamster	E	+	Li et al. (1980)
+	pHe	BP8 ascites	Murine sarcoma	E	−	Hofer and Mivechi (1980)
+	pHi	BP8 ascites	Murine sarcoma	E	−	Hofer and Mivechi (1980)
+	pHe	Fibrosarcoma	Mouse	E	+	Urano et al. (1980)
+	pHi&e	BP8 ascites	Murine sarcoma	E	+	Mivechi et al. (1981)
+	pHi&e	BP8 ascites	Murine sarcome	AH	+	Mivechi et al. (1981)
+	pHe	Glial	Human astrocytoma	E	+	Rottinger and Mendonca (1982)
+	pHe	M 8013S	Mouse mammary carcinoma	E	−	Haveman (1983)

[a] Data extracted from Wike-Hooley et al. (1984).
[b] (−), Protection; (+), Sensitization; (0), No-Effect.
[c] pHe, extracellular pH; pHi, intracellular pH.
[d] E, Euoxic; H, Hypoxic; CH, Chronically Hypoxic; AH, Acutely Hypoxic; EH, An inhibitor of oxidative phosphorylation added.

sensitivity due to pH reduction of the culture medium was demonstrated in all lines with the exception of the SDB rat mammary carcinoma (Dickson and Oswald 1976). In this study, however, damage was assessed by trypan blue exclusion, and time for re-initiation of cell division.

Several aspects of the *in vitro* pH sensitizing effect attest to its potential *in vivo* significance. The pH sensitizing effect is manifest over a pH range which is observed in tumor tissue, i.e. pH 6.6 to 7.0 (Gerweck 1977). This is illustrated in Fig. 3. Chinese hamster ovary (CHO) cells were exposed to a fixed treatment of 42 °C for 4 hours at extracellular pH values ranging from 7.6 to pH 6.7. Variation of pH between 7.6 and 7.1 does not significantly affect thermal sensitivity, however, sensitivity increases significantly as the medium pH decreases below pH 7.1. The magnitude of the pH sensitizing effect is most pronounced at temperatures which are only moderately lethal under 'normal' (pH 7.4) conditions (Gerweck 1977). This is illustrated in Fig. 4. At 41–42 °C, cells become refractory to heat inactivation after a 200-minute treatment at pH 7.4, however, at pH 6.7 this refractory response does not develop over the treatment time examined. At higher temperatures, the development of thermal resistance is not observed at normal pH, and the magnitude of pH sensitization is reduced. Figure 5 illustrates the temperature dependency of the pH enhancement effect on CHO and cultured human glioblastoma cells (Gerweck and Richards 1981). the 'pH enhancement ratio' is the ratio of the inactivation rates at pH 6.7 vs pH 7.4. Terminal slope

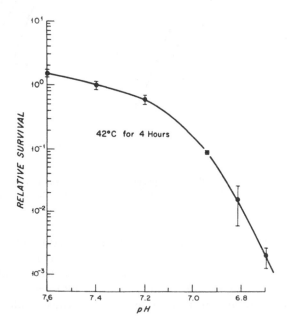

Figure 3: Influence of pH on the thermal sensitivity of Chinese hamster cells (Gerweck 1977). Data have been normalized to the surviving fraction of pH 7.4.

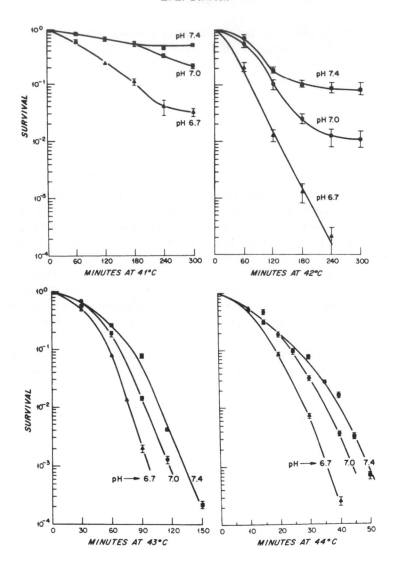

Figure 4: Chinese hamster ovary cells were exposed to heat treatments at temperatures ranging from 41 °C to 44 °C. The extracellular medium pH was 7.4, 7.0 or 6.7 during the hyperthermia treatments (Gerweck 1977).

ratios are plotted, except for the 42 °C data for CHO cells. Similar to the effect with CHO cells, pH sensitization of human glial cells is more pronounced in the lower temperature range. However, as the treatment temperature drops below 43 °C in glial cells and 42 °C in CHO cells, pH sensitization decreases. Further temperature decreases do not cause additional sensitization.

As is the case with single heat treatments, reduced pH also sensitizes cells to fractionated hyperthermia (Nielsen and Overgaard 1979; Gerweck *et al.*

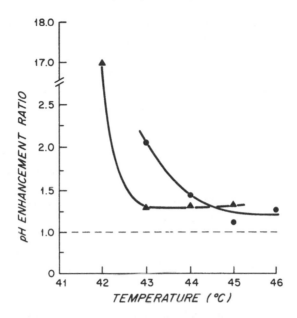

Figure 5: The pH enhancement ratio is plotted at various temperatures for human glioblastoma cells (circles) and Chinese hamster cells (triangles). The pH enhancement ratio is the ratio of the inactivation rates at pH 6.7 vs 7.4 at the indicated temperatures (Gerweck *et al.* 1981).

1980, 1983; Goldin and Leeper 1981). The potential clinical significance of this effect is unclear, however, as cells residing in a low pH environment would be least likely to survive an initial heat treatment.

The vast majority of *in vitro* studies have been performed under conditions where pH is rapidly altered, e.g., over 0 to 30 minutes, and this pH adjustment is then followed by heat treatment from a few minutes to 6 to 8 hours later. Under *in vivo* conditions, low pH presumably develops more gradually and cells may reside in an acidic environment for more prolonged periods. It is not known if cells adapt to low pH under these conditions, thereby reversing the pH sensitizing effect. Acute reduction in medium pH to 6.7, followed by a culturing at pH 6.7 for up to 5 days, did not appreciably influence the pH sensitizing effect to 42 °C in one study (Gerweck *et al.* 1982). However, prolonged maintenance of Chinese hamster HA1 cells at pH 6.8 prior to heating at pH 6.8 clearly leads to a decrease in their thermal sensitivity (Hahn and Shiu 1986). The method of controlling medium pH, i.e. by CO_2 adjustment versus variation in lactate concentration or sodium bicarbonate concentration, may have influenced the results obtained. Clearly, further studies are needed on various cell lines to determine if metabolic acidification and chronic acidity are important modifiers of the pH sensitizing effect.

It is not presently known if change in extracellular pH only is necessary for the pH sensitizing effect. In general, most studies have shown that when extracellular pH decreases below approximately 7.0, intracellular pH also decreases approximately 0.5 to 1.0 unit/per extracellular pH unit (reviewed in Waddell and Bates 1969; Roos and Boron, 1981). More recently, however, Gonzales-Mendez *et al.* (1982) reported that a decrease in extracellular pH from 7.5 to approximately 6.2 was virtually without effect on the internal pH of HA-1 cells. In view of the fact that HA-1 cells are sensitized by less pronounced extracellular pH decreases (Li *et al.* 1980), change in extracellular pH alone appears important in thermal sensitization. In apparent contrast to these results (constancy of intracellular pH), Gillies *et al.* (1982) observed a substantial change (0.9 units) in intracellular pH when extracellular pH is decreased from 7.5 to 6.2 in Ehrlich ascites cells (thermal sensitivity was not evaluated). In other studies, Hofer and Mivechi (1980) observed a more rapid release of [125]Iododeoxyuridine-labeled cells incubated with a H^+ ionopore, than in cells heated without ionopore. Haveman (1979) showed that an uncoupler of oxidative phosphorylation (carbonylcyanide 3-chlorphenyl hydrazone) sensitized cells to hyperthermia, especially above and below pH 8.0. These data suggest that the increased thermal sensitivity depended upon a decreased intracellular pH or to a pH-mediated decrease in cell energy status.

Resolution of the relationship between extracellular pH and intracellular pH, as well as whether intracellular acidification is necessary for thermal sensitization is of more than academic interest. For example, a method which holds promise for the non-invasive assessment of tissue pH, i.e. ^{31}P-NMR spectroscopy, is believed to primarily measure intracellular pH. If intracellular pH does not reflect extracellular pH, and exracellular pH acidification only is required for increased thermal sensitivity, then NMR measurements will not be useful for predicting thermal sensitivity based on pH considerations.

In summary, abrupt changes in extracellular pH for a few minutes to several hours before heat treatment, increases cellular thermal sensitivity. Whether this increased thermal sensitization is due to a decrease in intracellular and extracellular pH, has not been resolved.

E. OXYGEN CONCENTRATION AND THERMAL SENSITIVITY

Substantial evidence, both direct and indirect, demonstrates or suggests the presence of hypoxic foci in rodent and human tumor tissue. Tumor oxygen tension, tumor vasculature, and their interrelationship are extensively discussed by Vaupel (1979). Several studies have compared the response of cells to hyperthermia under oxic and hypoxic conditions over the past 10 years, e.g. Gerweck *et al.* (1974); Harisiadis *et al.* (1975); Kim *et al.* (1975); Overgaard and Bichel (1977); Power and Harris (1977); Bass *et al.* (1978). Varying results have been obtained. In general, it appears that cells are equally or more sensitive to hyperthermia under hypoxic compared to oxic conditions. Differences between the results are likely explained by factors such as duration of hypoxia prior to heating, pH differences, and differences in cell density. In studies with CHO cells (Gerweck *et al.* 1981), short-term variation in oxygen concentration prior to and during heat treatment did not influence cellular response to hyperthermia. However, when CHO cells were maintained under prolonged hypoxic conditions their sensitivity substantially increased (Fig. 6). This increased thermal sensitivity was largely but not entirely due to acidification of the medium due to cell metabolism as summarized in Table 3. Other possible contributing mechanisms to hypoxic sensitization of cells include decreasing cell energy status, as will be discussed later.

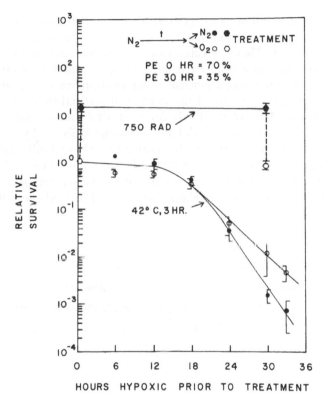

Figure 6: Chinese hamster ovary cells were maintained under hypoxic (N_2) conditions for up to 33 hours and then exposed to a 42 °C heat treatment for 3 hours. Cells were heated under hypoxic conditions (●) or reoxygenated immeidately prior to treatment (○). The response of cells exposed to ionizing radiation under the same conditions is also indicated (Gerweck *et al.* unpublished studies).

Table 3.
Influence of pH on the heat sensitivity of chronically hypoxic cells[a]

Treatment condition	pH ± 0.05	Surviving fraction after 42 °C for 3 hours
Acute oxygenated or hypoxia	7.38	0.125 ± 0.023
Chronic hypoxia	6.85	0.0085 ± 0.0024
Chronic hypoxia (pH adjusted)	7.31	0.049 ± 0.018

[a] Approximately 5×10^5 cells/ml were heated under oxygenated or acute (3.5 hr) hypoxic conditions, or following 27 hours of culturing under hypoxic conditions. In the pH adjustment studies, the pH was adjusted immediately prior to heat treatment (Gerweck *et al.* 1979).

In summary, short-term oxygen deprivation is without significant effect on cellular thermal sensitivity under optimal and constant nutrient and pH conditions. However, culturing conditions which give rise to changes in pH and/or nutrient status also increase thermal sensitivity.

F.   GLUCOSE AND THERMAL SENSITIVITY

Several studies [e.g. Gullino *et al.* (1967); Shapot and Blinov (1974); Sauer *et al.* (1982)] have demonstrated and quantitated the high glycolytic rate of several rodent tumors. These elevated rates of glucose consumption lead to low glucose levels in tumor interstitial fluid. Burgess and Sylven (1962) and Gullino *et al.* (1965) measured glucose concentrations in tumor interstitial fluid in the range of 0.1 to 0.01 of the normal interstitial fluid range.

In *in vitro* studies, Kim *et al.* (1980) demonstrated that reduced levels of oxygen alone, or glucose alone, did not sensitize cells to hyperthermia. However, when cells were exposed to reduced levels of both oxygen and glucose, thermal sensitivity increased markedly, as shown in Fig. 7. Similar results have been obtained with the use of inhibitors of glycolysis or respiration (Song *et al.* 1979; Haveman and Hahn 1981; Laval and Michel 1982; and Nagle *et al.* 1982).

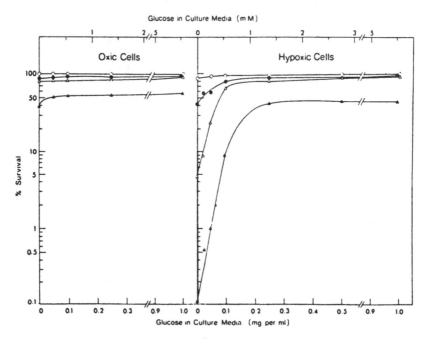

Figure 7: Survival of oxic and hypoxic HeLa cells exposed to different temperatures for 2 hours with various concentrations of glucose in culture media. Cell survival is expressed as a percentage of the unheated control cells. Plating efficiency of control cells was 60–70 per cent. Symbols for oxygen concentrations: ▲, 0 per cent. △, 0.5 per cent. ●, 1 per cent. ○, 21 per cent (Kim *et al.* 1980).

G. ENERGY STATUS AND THERMAL SENSITIVITY

Under conditions of extreme nutrient deprivation, cell death occurs rapidly at 37 °C. Nutrient deprivation (both oxygen and glucose) likely give rise to the necrotic foci (Hardman 1940; Thomlinson and Gray 1955) which are observed in tumors as the radial distance from functioning capillaries increases. Between the nutrient-rich capillaries and viable zone, cells reside in a gradient of decreasing oxygen and glucose concentration. In CHO, and probably other cells, toxicity resulting from oxygen and glucose deprivation is markedly more toxic at 37 °C (or elevated temperatures), than deprivation of either substrate alone (Gerweck *et al.* 1984). Oxygen and glucose (and probably to a lesser extent amino acids and fatty acids under aerobic conditions) are major substrates for ATP synthesis. Work by Laval and Michel (1982) and Nagle *et al.* (1982) provides evidence for a key role of this metabolite in thermal sensitization. These investigators showed a decreased ATP level in cells which were rendered heat sensitive by treatment with inhibitors of ATP synthesis. Figure 8 demonstrates a strong correlation between cell thermal sensitivity and cellular ATP levels (Gerweck *et al.* 1984). Little information is available regarding the overall energy level or oxygen distribution of cells within tumors. However, these *in vitro* studies suggest that hyperthermia will be toxic to energy-deprived tumor cell populations.

H. SUMMARY

Tumors exhibit high rates of aerobic and anaerobic glycolysis compared to most normal tissues. The resultant high rates of acid production, combined with a normal or reduced capacity to clear the produced acid, results in tissue acidosis. *In vitro* studies show that reduction in extracellular pH to values similar to those observed in tumors increases the hyperthermic sensitivity of cells. The relationship between extracellular pH, intracellular pH, and thermal sensitization, however, is not fully understood. Neither has the relationship between a tumor's elevated glycolytic rate, glucose availability, and its energy status been evaluated under aerobic conditions. However, the high tumor demand for glucose, in combination with irregular tumor vasculature, likely leads to a reduced availability of glucose and oxygen, and reduces ATP production via glycolysis. Results of *in vitro* studies demonstrate that cellular ATP depletion (via nutrient reduction) sensitizes cells to hyperthermia. Although glycolysis plays a major role in the development of tissue acidosis and in ATP production, local tissue perfusion also plays a pivotal role in thermal sensitization by supplying and clearing critical metabolites. Not all areas of tumors are uniformly acidic, nor is it likely that all cells in tumors are equally energy deprived. For this reason, hyperthermia will likely be most effective for cancer therapy when used in conjunction with other treatment modalities whose effects are complementary.

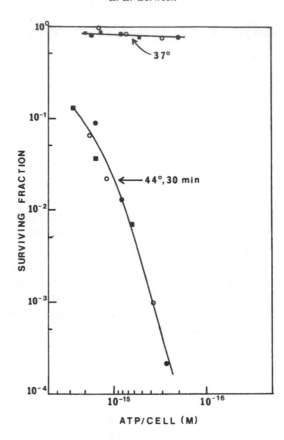

Figure 8: The fractional survival of Chinese hamster ovary cells is plotted as a function of cellular ATP concentration. Cellular ATP levels were varied by varying the nutrient medium oxygen and glucose concentration (data extracted from Gerweck *et al.* 1984).

NOTE IN PROOF

Chu and Dewey have recently reported that the relationship between thermal sensitivity and pH most strongly correlates with intracellular rather than extracellular pH. These results were observed in both normal and pH adapted cells. Manuscript in press (*Radiat. Res.* **114**, 1988).

REFERENCES

Aisenberg, A.C. (1961). *The Glycolysis and Respiration of Tumors.* Academic Press, Inc., New York and London.

Bass, H., Moore, J.L. and Coakley, W.T. (1978). Lethality in mammalian cells due to hyperthermia under oxic and hypoxic conditions. *Int. J. Radiat. Biol.* **33**, 57–66.

Burgess, E.A. and Sylven, B. (1962). Changes in glucose and lactate content of ascites fluid in blood plasma during growth and decay of ELD ascites tumor. *Br. J. Cancer* **16**, 298–305.

Burk, D. (1939). Cold Spring Harbor Symposium, *Quant. Biol.* **7**, 420–436.

Calderwood, S.A. and Dickson, J.A. (1983). pH and tumor response to hyperthermia. *Adv. Radiat. Biol.* **10**, 135–190.

Dickson, J.R. and Calderwood, S.K. (1983). Thermosensitivity of neoplastic tissue *in vivo*. In *Hyperthermia in Cancer Therapy*, F.K. Storm (Ed.). Boston: G.K. Hall, pp. 63–140.

Dickson, J.A. and Oswald, B.E. (1976). The sensitivity of a malignant cell line to hyperthermia (42 °C) at low intracellular pH. *Br. J. Cancer* **34**, 262–271.

Freeman, M.L., Dewey, W.C. and Hopwood, L.E. (1977). Effect of pH on hyperthermic cell survival. *J. Natl. Cancer Inst.* **58**, 1837–1839.

Freeman, M.L., Raaphorst, G.P., Hopwood, L.E. and Dewey, W.C. (1980). The effect of pH on cell lethality induced by hyperthermic treatment. *Cancer* **45**, 2291–2300.

Freeman, M.L., Boone, M.L.M., Ensley, B.A., and Gillette, E.L. (1981). The influence of environmental pH on the interaction and repair of heat and radiation damage. *Int. J. Radiat. Oncol. Biol. Phys.* **7**, 761–764.

Gerweck, L.E., Gillette, E.L. and Dewey, W.C. (1974). Killing of Chinese hamster cells *in vitro* by heating under hypoxic or aerobic conditions. *Eur. J. Cancer* **10**, 691–693.

Gerweck, L.E. (1977). Modification of cell lethality at elevated temperatures. The pH effect. *Radiat. Res.* **70**, 224–235.

Gerweck, L.E., Nygaard, T.G., and Burlett, M. (1979). Response of cells to hyperthermia under acute and chronic hypoxic conditions. *Cancer Res.* **39**, 966–972.

Gerweck, L.E., Jennings, M. and Richards, B. (1980). Influence of pH on the response of cells to single and split doses of hyperthermia. *Cancer Res.* **40**, 4019–4024.

Gerweck, L.E. and Richards, B. (1981). Influence of pH on the thermal sensitivity of cultured human glioblastoma cells. *Cancer Res.* **41**, 845–849.

Gerweck, L.E., Richards, B. and Jennings, M. (1981). The influence of variable oxygen concentration on the response of cells to heat or X-irradiation. *Radiat. Res.* **85**, 314–320.

Gerweck, L.E., Richards, B. and Michaels, H.B. (1982). Influence of low pH on the development and decay of 42 °C thermotolerance in CHO cells. *Int. J. Radiat. Oncol. Biol. Phys.* **8**, 1935–1941.

Gerweck, L.E., Dahlberg, W.K., Akonuman, A. and Sarenji, E. (1983). Effect of pH on single and fractionated heat treatments at 42–45 °C. *Cancer Res.* **43**, 1163–1167.

Gerweck, L.E., Dahlberg, W.C., Epstein, L. and Shimm, D.S. (1984). Influence of nutrient and energy deprivation on cellular response to single and fractionated heat treatments. *Radiat. Res.* **99**, 573–581.

Gerweck, L.E. (1985). Hyperthermia in cancer therapy: The biological basis and unresolved questions. *Cancer Res.* **45**, 3408–3414.

Gillies, R.J., Ogino, R., Shulman, R.G. and Ward, D.C. (1982). ^{31}P-NMR evidence for the regulation of intracellular pH by Ehrlich ascites tumor cells. *J. Cell Biol.* **95**, 24–28.

Goldin, E.M. and Leeper, D.B. (1981). The effect of reduced pH on the induction of thermotolerance. *Radiology* **141**, 505–508.

Gonzales-Mendez, R., Wemmer, D., Hahn, D., Wade-Jardetzki, N. and Jardetzki, O. (1982). Continuous flow NMR culture system for mammalian cells. *Biochim. Biophys. Acta*, **720**, 274–280.

Gullino, P.M., Grantham, F.H., Smith, S.H. and Haggerty, A.C. (1965). Modification of the acid-base status of the internal milieu of tumors. *J. Natl. Cancer Inst.* **34**, 857–869.

Gullino, P.M., Grantham, F.H. and Courtney, A.H. (1967). Glucose consumption by transplanted tumors *in vivo*. *Cancer Res.* **27**, 1031–1040.

Hahn, G.M. and Shiu, E.C. (1986). Adaptation to low pH modifies thermal and thermo-chemical responses of mammalian cells. *Int. J. Hyperthermia*, **2**, 379–387.

Hardman, J. (1940). The angioarchitecture of the gliomata. *Brain* **63**, 91–118.

Harisiadis, L., Hall, E.J., Kraijevic, U. and Borek, C. (1975). Hyperthermia: Biological studies at the cellular level. *Radiology* **117**, 447–452.

Haveman, J. (1979). The pH of the cytoplasm as an important factor in the survival of *in vitro* cultured malignant cells. Effects of Carbonylcyanide 3-chlorophenylhydrazone. *Eur. J. Cancer* **15**, 1281–1288.

Haveman, J. and Hahn, G.M. (1981). The role of energy in hyperthermia-induced mammalian cell inactivation: A study of the effects of glucose starvation and an uncoupler of oxidative phosphorylation. *J. Cell Physiol.* **107**, 237–241.

Haveman, J. (1983). Influence of pH and thermotolerance on the enhancement of X-ray induced inactivation of cultured mammalian cells by hyperthermia. *Int. J. Radiat. Biol.* **43**, 281–289.

Hofer, K.G. and Mivechi, N.F. (1980). Tumor cell sensitivity to hyperthermia as a function of extracellular and intracellular pH. *J. Natl. Cancer Inst.* **65**, 621–625.

Jahde, E. and Rajewsky, M.F. (1983). Sensitization of clonogenic malignant cells to hyperthermia by glucose-mediated tumor-selective pH reduction. *J. Cancer Res. Clin. Oncol.* **104**, 23–30.

Kim, S.H., Kim, J.H. and Hahn, E.W. (1975). Enhanced killing of hypoxic tumor cells by hyperthermia. *Br. J. Radiol.* **48**, 872–874.

Kim, S.H., Kim, J.H., Hahn, E.W. and Ensign, N.A. (1980). Selective killing of glucose and oxygen-deprived HeLa cells by hyperthermia. *Cancer Res.* **40**, 3459–3462.

Laval, F. and Michel, S. (1982). Enhancement of hyperthermia-induced cytoxicity upon ATP deprivation. *Cancer Letter* **15**, 61–65.

Li, G.C., Shiu, E.C. and Hahn, G.M. (1980). Recovery of cells from heat-induced potentially lethal damage: Effects of pH and nutrient environment. *Int. J. Radiat. Oncol. Biol. Phys.* **6**, 577–582.

Meyer, K.R., Hopwood, L.E. and Gillette, E.L. (1979). The thermal response of mouse adenocarcinoma cells at low pH. *Eur. J. Cancer* **15**, 1219–1222.

Mivechi, N.F., Hofer, K.G. and Hofer, M.G. (1981). Influence of hypoxia and acidity on thermal radiosensitization and direct heat-induced death of BP-8 sarcoma cells. *Radiology* **138**, 465–471.

Nagle, W.A., Moss, A.J. and Baker, M.L. (1982). Increased lethality from hyperthermia at 42 °C for hypoxic Chinese hamster cells heated under conditions of energy deprivation. *Natl. Cancer Inst. Monogr.* **61**, 107–110.

Nielsen, O.S. and Overgaard, J. (1979). Effect of extracellular pH on thermotolerance and recovery of hyperthermia damage *in vitro*. *Cancer Res.* **39**, 2772–2778.

Overgaard, J. (1976). Influence of extracellular pH on the viability and morphology of tumour cells exposed to hyperthermia. *J. Natl. Cancer Inst.* **56**, 1243–1250.

Overgaard, J. and Bichel, P. (1977). Hyperthermic effect on exponential and plateau ascites tumor cells *in vitro* dependent on environmental pH. *Radiat. Res.* **70**, 449–454.

Power, J.A. and Harris, J.W. (1977). Response of extremely hypoxic cells to hyperthermia: Survival and oxygen enhancement ratios. *Radiology* **123**, 767–770.

Roos, A. and Boron, W.F. (1981). Intracellular pH. *Physiol. Rev.* **6**, 296–434.

Rottinger, E.M. and Mendonca, M. (1982). Radioresistance secondary to low pH in human glial cells and Chinese hamster ovary cells. *Int. J. Radiat. Oncol. Biol. Phys.* **8**, 1309–1314.

Sauer, L.A., Stayman, J.W. and Dauchy, R.T. (1982). Amino acid, glucose and lactic acid utilization *in vivo* by rat tumors. *Cancer Res.* **42**, 4090–4097.

Shapot, V.S. and Blinov, V.A. (1974). Blood glucose levels and gluconeogenesis in animals bearing transplantable tumors. *Cancer Res.* **34**, 1827–1832.

Song, C.W., Guertin, D.P. and Levitt, S.M. (1979). Potentiation of cytotoxicity of 5-thio-D-glucose on hypoxic cells by hyperthermia. *Int. J. Radiat. Oncol. Biol. Phys.* **5**, 965–970.

Thistlewaite, A.J., Leeper, D.B., Moylan, D.J. and Nerlinger, R.E. (1985). pH distribution in human tumors. *Int. J. Radiat. Oncol. Biol. Phys.* **11**, 1647–1652.

Thomlinson, R.H. and Gray, L.H. (1955). The histological structure of some human lung cancers and their possible implications for radiotherapy. *Br. J. Cancer* **9**, 539–549.

Urano, M., Gerweck, L.E., Epstein, R., Cunningham, M. and Suit, H.D. (1980). Response of a spontaneous murine tumor to hyperthermia: Factors which modify the thermal response *in vivo*. *Radiat. Res.* **83**, 312–322.

Urano, M., Montoya, V. and Booth, A. (1983). Effect of hyperglycemia on the thermal response of murine normal and tumor tissue. *Cancer Res.* **43**, 453–455.

Vaupel, P. (1979). Oxygen supply to malignant tumors. In *Tumor Blood Circulation: Angiogenesis, Vascular Morphology and Blood Flow of Experimental and Human Tumors*, H.I. Peterson (Ed.), Boca Raton, Florida: CRC Press, pp. 143–168.

Vaupel, P., Frinak, S. and Bicher, H.I. (1981). Heterogenous oxygen partial pressure and pH distribution in C3H mouse mammary Adenocarcinoma. *Cancer Res.* **41**, 2008–2013.

Von Ardenne, M. (1972). Selective multiphase cancer therapy: Conceptual aspects and experimental basis. *Adv. Pharmacol. Chemoth.* **10**, 339–380.

Waddell, W.J. and Bates, R.G. (1969). Intracellular pH. *Physiol. Rev.* **49**, 285–329.

Warburg, O. (1926). Uber den stoffwechsel der Fumoren, Berlin: Springer. Translated: *The Metabolism of Tumors*. London: Arnold Constable, 1930.

Weinhouse, S. (1976). The Warburg hypothesis fifty years later. *Z. Krebsforsch.* **87**, 115–126.

Wike-Hooley, J.L., Haveman, J. and Reinhold, H.S. (1984). The relevance of tumor pH to the treatment of malignant disease. *Radiother. Oncol.* **2**, 343–366.

Hyperthermia and Oncology, Vol. 1, pp. 99–119 (1988)
Urano and Double (Eds)

Chapter 4.2

Modification of thermal effects: chemical modifiers

JAE HO KIM
*Department of Radiation Oncology, Memorial Sloan-Kettering Cancer Center,
New York, NY 10021, USA*

A. INTRODUCTION

It is well established that exposure to elevated temperatures can produce regression of cancer in animal models and in humans (Hahn 1982); however, the physiological mechanisms involved are somewhat undefined, and the clinical therapeutic index remains low. In pursuing the cellular mechanisms of heat-induced cytotoxicity, early cell culture studies suggested that cancer cells were more sensitive to heat than were normal cells (Chen and Heidelberger 1969; Giovanella *et al.* 1976). However, more recent experimental data tend to indicate that there may be very little inherent difference in the intrinsic heat sensitivity between the transformed cell and its normal cell counterpart (Hahn 1980). On the other hand, there now exists clear evidence that the thermosensitivity of mammalian cells can be influenced by various environmental factors of cell culture, such as pH, glucose, oxygen, and nutritional status (Hahn 1974; Dewey *et al.* 1977; Gerweck 1977; Kim *et al.* 1978).

In addition to the environmental factors, several chemical agents whose mode of action is reasonably well understood in cells at 37 °C have been shown to selectively modify the thermosensitivity of cells; findings from such studies have provided important clues toward the identification of the principal mechanism of hyperthermia-induced cytotoxicity. There is also evidence in the literature that the use of such chemical agents may enhance the beneficial effects of hyperthermia alone or in combination with conventional radiotherapy and chemotherapy in the treatment of cancer. This paper on chemical modifiers reviews the present state of the art.

B. HYPERTHERMIC SENSITIZERS

Hyperthermic sensitizer can be defined as any compound that is not cytotoxic at 37 °C but at elevated temperatures becomes cytotoxic. So characterized, this would exclude most of the chemotherapeutic agents from the present discussion, since they are by themselves cytotoxic at 37 °C, but when combined with heat, further enhance cellular injury (e.g., cis-platinum, nitrosourea). There are at least two classes of such agents that enhance the lethal effects of heat. Agents that potentiate hyperthermic cytotoxicity by lowering temperature threshold include local anesthetics, short chain aliphatic alcohols, energy depleters, and lactate transport inhibitors. The other class of agents includes polyamines and thiol containing drugs which become cytotoxic at temperatures of 42.5 °C or higher. Some heat sensitizers (e.g., membrane active agents and glycolytic inhibitors) also exhibit to enhance the radiation effect at 37 °C, although the cellular mechanism of potentiating the effects of heat or radiation are not commonly shared with. Since the pharmacology of several heat sensitizers is well known, clinical experience with them is extensive and they are readily localized in tissues, these factors may favor their prompt clinical application, if pre-clinical studies of drugs with well defined *in vivo* tumor systems show an improved therapeutic gain. Some heat sensitizers are now undergoing phase I and II clinical trials (Robins et al. 1985).

1.   Membrane active agents
The importance of the membrane as a critical target in the enhancement of heat-induced cell killing has often been emphasized (Yatvin 1977; Wallach 1978; Yau 1979; Dewey et al. 1980; Hahn 1982). Hence, agents that modify the membrane activity have been shown to interact synergistically with heat in a variety of cellular systems. Anesthetics, tranquilizers, as well as other agents such as short-chained aliphatic alcohols are known to interact with the cell membrane, causing alterations in their structural and functional organization (Seeman 1972; Sheetz and Singer 1974; Silva ert al. 1979).

Local anesthetics belong to a class of clinically useful compounds that exert their pharmacological effect by blocking the generation and the conduction of the nerve impulse. Their main site of action is regarded to be the cell membrane. Thus, local anesthetics have been employed in many cellular studies to modify membrane-mediated cellular processes and regulatory mechanisms. In a series of experiments aimed at establishing the importance of membrane fluidity and thermosensitivity of cells, Yatvin et al. (1977) have shown that procaine–HCl increased hyperthermic killing of an unsaturated fatty acid auxotroph of *Escherichia coli* and the thermosensitivity of bacterial cells was proportional to embrane microviscosity (Dennis and Yatvin 1981).

In mammalian cells in culture, Yau first showed that procaine-HCl enhanced the cytotoxicity of heat at 43 °C in mouse L5178Y lymphoma cells (Yau 1979). The potentiating effect was dependent on the temperature and concentration of drugs (Fig. 1). The potentiation was evident at minimum drug concentration of 1.0 mM which can be readily achievable in humans. Subsequent studies showed that dibucaine, one of the longest-acting anesthetics, was the most effective and procaine the least effective potentiator on a molar basis among the local anesthetics screened (Yau and Kim 1980). The thermal sensitization of cells by procaine occurs equally well for cells heated and

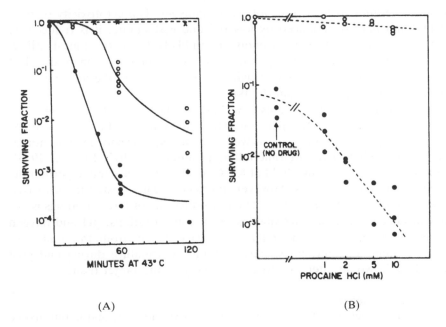

(A)   (B)

Figure 1: Cell surviving fraction of L5178Y lymphoma cells as a function of time after exposure to 43 °C X: cells exposed to 37 °C; ○: cells exposed to heat only; ●: cells exposed to heat plus procaine (10 mM). (Panel A). Dose dependency of procaine in potentiating heat killing (43 °C for 60 min.). ○: cells at 37 °C; ●: cells exposed to 43 °C (Panel B) (from Yau 1979).

exposed to the drug either G_1 or S-phase, normally the most heat-resistant and sensitive phase of the cell cycle, respectively (Coss and Dewey 1982). Unlike the results obtained with procaryotes, there was no direct correlation between the membrane fluidity and enhanced thermosensitivity of mammalian cells (Lepock 1982).

The effectiveness of local anesthetics as a heat sensitizer in an *in vivo* tumor system was first carried out with lidocaine infusion of CA 755 mammary adenocarcinoma growing in BDF_1 mice. There was a significant increase in the animals' survival when combined with heating for 1 hour in a 43.5 °C water bath (Yatvin *et al.* 1979). Subsequent study with lidocaine administered through intraperitoneal route also resulted in a significant increase in the tumor control relative to that of heat alone (Robins *et al.* 1982). In view of the fact that extensive clinical experience with local anesthetics and pharmacokinetics of the drug is well known and most local anesthetics interact synergistically with radiation, this class of drugs needs to be further investigated in a variety of *in vivo* tumor systems.

Short chain aliphatic alcohols including ethanol were found to be an excellent heat sensitizer (Li and Hahn 1978). Ethanol exhibits to lower temperature threshold. Thus, one degree of heat is about equivalent 1 per cent of ethanol in the medium (Li *et al.* 1980). The resemblance of the action of ethanol on mammalian cells is so close to that of heat that ethanol is labeled as a heat analog. For example, exposure of cells to heat or ethanol induces tolerance to heat, ethanol, or adriamycin. The cytotoxicity of both ethanol and heat is enhanced by low pH and cysteamine; it is reduced by deuterium oxide. However, alcohols are unlikely to be of clinical utility, since the amount of alcohol required would be too high for human use. But the study with such agents may provide some insight into the mechanism of action of heat itself (Hahn and Li 1982).

Other membrane active agents have been studied as heat sensitizers. For example, amphotericin B, a polyene antibiotic, anti-fungal agent that binds to sterols of plasma membrane, as well as some organic solvents potentiates mammalian cells to hyperthermic killing (Hahn et al. 1977; Li et al. 1978). There is a temperature threshold of 42 °C to 43 °C to cause the hyperthermic potentiation. The drug was, however, ineffective when tested in an in vivo tumor system (Hahn et al. 1977). The reason for the different findings is not known. Of course, pharmacokinetics, drug concentrations, pH and oxygen concentration in tumors of animals can vary widely from those in vitro culture system. Further, physiology of tumor vasculature during and post hyperthermia would alter the drug availability to the target tissue.

2.   Energy depleters

A relationship between temperatures, energy expenditure, and maintenance of cellular integrity under adverse conditions have been recognized for decades. The protective effect of the hypothermic state on survival of brain and cardiac function under hypoxic conditions is widely utilized. Applying the energy equilibrium concept to hyperthermic conditions, Kim and others demonstrated that the capacity to continue energy production by either oxidative or glycolytic metabolism is a prerequisite for cellular survival at elevated temperatures (Kim et al. 1980; Haveman and Hahn 1981; Gerweck et al. 1984). Thus, hypoxic cells were not dramatically more sensitive to heat unless they were also deprived of glycolyzable substrates. Similarly, heated glucose-deprived cells could survive relatively well on fatty acids and endogenous substrates unless they were also deprived of oxygen. When cells are heated under conditions of combined hypoxia and glucose deprivation, the rate of cell kill is dramatically increased.

The above findings are generally consistent with the following concept. Temperature elevation increases the energy expenditure of the cell, but the cell can compensate, to a degree, by increasing energy production. However, if, by reason of substrate depletion, drug effects an alteration of the internal metabolic condition, either (a) energy production is reduced, or (b) energy expenditures are further increased, and the heated cell can become relatively energy depleted and less able to maintain its metabolic or structural integrity. In a time-dependent fashion, this can lead to loss of the clonogenic capacity. To support the foregoing concept, studies with energy depleters have shown that the deprivation of cellular energy by means of either inhibition of glycolysis or oxidative phosphorylation enhances the thermosensitivity of either oxic or hypoxic cells.

(i) Glycolytic inhibitors. Among inhibitors of glycolysis, 5-thio-D-glucose, a nearest analogue of D-glucose, was initially shown to be preferentially cytotoxic against hypoxic tumor cells in culture (Song *et al.* 1977). Kim *et al.* (1978) subsequently showed that 5-thio-D-glucose selectively enhanced the thermosensitivity of hypoxic HeLa cells at temperatures as low as 41 °C (Fig. 2). More recently, Nagle *et al.* (1985) obtained a similar result with another glycolytic inhibitor, pentalenolactone, a specific inhibitor of glyceral-dehydephosphate dehydrogenase. The selective thermosensitization of hypoxic cells by pentalenolactone is again qualitatively similar with the previous result obtained with sodium fluoride (Nagle *et al.* 1982). In this experiment, cells become selectively heat sensitive only under hypoxic conditions, since cells under hypoxia are solely dependent upon their source of energy via anaerobic glycolysis. The thermosensitivity of oxic cells is not influenced by the drugs.

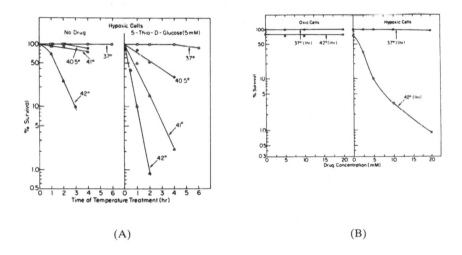

(A)   (B)

Figure 2: Per cent cell survival of HeLa cells exposed to different temperatures with or without 5-thio-D-glucose (5 mM) under hypoxic conditions (Panel A). Survival of oxic and hypoxic HeLa cells exposed to 42 °C for 1 hour with varying concentrations of 5-thio-D-glucose (Panel B) (from Kim *et al.* 1978).

In pursuing the cellular mechanism of hypoxic cell thermosensitization by the glycolytic inhibitors, Kim *et al.* (1984) identified several other glycolytic inhibitors to be equally good heat sensitizers of both oxic and hypoxic cells. Lonidamine, 1-(2,4-dichlorophenyl)-1H-indazole-3-carboxylic acid, is a potent inhibitor of spermatogenesis in various mammalian species and posses-ses embryotoxic and anti-tumor effects (Silvestrini *et al.* 1983). Biochemical studies appeared to indicate that the major glycolytic inhibition was mediated through effects on the mitochondrial bound hexokinase (Floridi *et al.* 1981). Since 5-thio-D-glucose is a potent antifertility agent and hyperthermic sen-sitizer of hypoxic cells, Kim *et al.* (1984) carried out cell culture studies to determine whether lonidamine enhances the heat induced cytotoxicity. The observations of Kim *et al*. (1984) and Silvestrini *et al*. (1983) demonstrated lonidamine to be a remarkably effective hyperthermic sensitizer for cells both *in vitro* and *in vivo*. The hyperthermic sensitizing effect of lonidamine was greatly dependent upon the acidity of culture medium (Fig. 3); since an acidic pH is known to exist in the interstitial space of solid tumors, the pH dependency of the drug's effects may provide therapeutic selectivity in tumors *in vivo*. The experiments with an *in vivo* tumor system further showed that the combined dose of lonidamine (100 mg/kg i.p.) with local hyperthermia increased the tumor control rate up to 84 per cent, while heat treatment alone resulted in only 45 per cent tumor control (Kim *et al*. 1984).

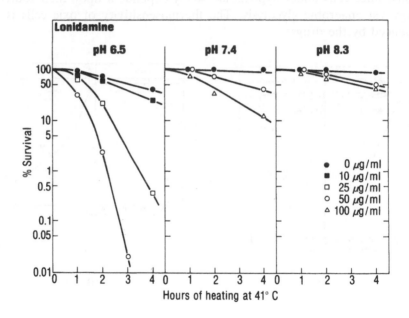

Figure 3: Changes in the per cent survival as a function of lonidamine exposure at 41 °C under three different pH conditions. The marked potentiation of the cytotoxicity by the drug (50 μg/ml) and heat is apparent under pH 6.5 (from Kim *et al*. 1984).

Gossypol, a polyphenolic aldehyde extracted from cotton plants, is another potent antifertility agent in man (Kalla 1982). Biochemical studies of gossypol have shown that the principal action of the drug is an inhibition of glycolytic and mitochondrial bound enzymes and interference of ion transport (Abou-Donia 1976; Hong *et al*. 1983). Gossypol was not cytotoxic at 37 °C (10 μg/ml). When HeLa cells were exposed to gossypol at 41 and 42 °C, significant potentiaion of hyperthermia induced cytotoxicity was obtained (Kim *et al*. 1985). The hyperthermic sensitizing effect of gossypol was again increased by an acidic pH and glucose deprivation.

Similar findings of the pH dependence for the drug's effect are reported with both lonidamine and quercetin, a lactate transport inhibitor. The hyperthermic sensitization of drugs under low pH can be understood in the context of the energy equilibrium of the cell as discussed in an earlier section. The accelerated metabolic demands of the hyperthermic state cannot be adequately met if the cell is de-energized accordingly. The rate of cell kill is increased under the conditions which produce an energy-depleted state. An acidic medium pH increases cellular energy demands, as the cell increases proton efflux through homeostatic mechanisms. A failure to restore the pH will further reduce the intracellular energy level by inhibiting glycolysis. In this context, it is conceivable that the treatment of cells by gossypol impairs cellular energy production through uncoupling of oxidative phosphorylation (Abou-Donia *et al*. 1976).

(ii) Inhibitors of oxidative phosphorylation. Since the main source of ATP production of oxic cells is both from oxidative phosphorylation and aerobic glycolysis, several studies have been carried out with mitochondrial inhibitors of oxidative phosphorylation to determine the influence of such inhibition on the cellular thermosensitivity. Uncouplers of electron transfer chains such as carbonyl cyanide chlorophenyl hydrazone (CCCP) in the presence of oxygen but in the absence of glucose significantly enhance hyperthermia cytotoxicity (Haveman and Hahn 1981; Laval and Michel 1982). The hyperthermic sensitization was most pronounced under acidic pH in the absence of glucose following exposure of cells to CCCP. The data again indicate that the availability of energy in cells before exposure to heat plays an important role in the response of these cells to heat.

Rhodamine 123 is a cationic fluorescent dye that binds specifically to mitochondria of living cells (Chen *et al*. 1982; Darzynkiewicz *et al*. 1982). It has been used as a supravital mitochondrial probe for long-term cell culture studies. The continuous exposure of cells to rhodamine 123 at high doses, however, inhibits oxidative phosphorylation, arrests cells in G_1 phase and induces a loss of reproductive capacity (Bernal *et al*. 1982). Further studies of the cytotoxic effects of rhodamine 123 with a variety of cell lines in culture have shown that the dye may be selectively cytotoxic against carcinoma cells, probably as a result of its prolonged retention in carcinoma mitochondria (Lampidis *et al*. 1983).

Using rhodamine 123, Kim *et al*. (1985) have recently shown that there was a pronounced enhancement of cell kill in glucose-deprived cells following exposure of HeLa cells to the drug (10 μg/ml) at 42 °C (Fig. 4). No enhanced effects of heat are seen in the glucose-fed cells. It is interesting to note that hyperthermic treatment under the acidic pH did not further enhance the cytotoxic effects of heat in the glucose-deprived cells; instead, the cell kill under the influence of reduced pH was far less than the cytotoxicity observed at pH 7.4, most probably as a result of reduced binding of the drug to the motochondria, since low pH may conceivably reduce the proton gradient across the mitochondrial membrane. Again, selective enhancement of hyperthermic cytotoxicity in glucose-deprived cells by rhodamine 123 appears to be in agreement with the foregoing concept of energy equilibrium that cell kill by heat is dependent on the critical level of cellular energy equilibrium.

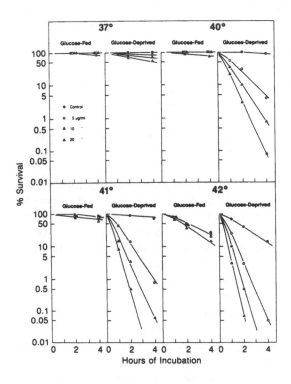

Figure 4: Percentage of cell survival of HeLa cells incubated with various concentrations of rhodamine 123 at 37 ° to 42 °C for 0 to 4 hours in the presence or absence of glucose. Cell survival is expressed as a percentage of unheated controls. Rhodamine 123 selectively increases the cytotoxic effects of hyperthermia in glucose-deprived cells relative to glucose-fed cells (from Kim *et al*. 1985).

3.   Inhibitors of lactate transport

It has been known for a long time that one of the consistent features of biochemical activity of neoplastic cells is the high aerobic glycolysis. The continued glycolysis of tumor cells results in copious production of lactate, which is readily excreted to the extracellular compartment. Increased aerobic and anaerobic glycolysis and the impairment of the exchange of metabolic products increase the hydrogen ion concentration in the extracellular space of the tumor tissue.

The transport of lactate across the plasma membrane of mammalian cells takes place via a specific transport system and not by simple diffusion of ions. Dubinsky and Racker (1978) and others have characterized the excretions of lactate as a proton-lactate symport mechanism (Spencer and Lehninger 1976). Belt *et al*. (1979) have identified two varieties of inhibitors of lactate transport which produced intracellular acidification and inhibition of glycolysis in Ehrlich ascites tumor cells. The documented inhibitors of lactate transport include cetain bioflavonoids and isobutyl carbonyl lactyl anhydride (iBCLA) (Johnson *et al.* 1980).

Quercetin, a bioflavonoid that produces lactate transport inhibition, is not cytotoxic at 37 °C (0.1mM). When HeLa cells were exposed to quercetin at 41 ° and 42 °C, significant potentiation of heat-induced cytotoxicity was observed (Kim *et al.* 1984). Treatment of cells with rutin, a structurally related bioflavonoids that lacks the property of lactate transport inhibition, showed no hyperthermic potentiation. The magnitude of the potentiation was dependent on the drug concentration, pH of the culture medium, temperature and duration of treatment. The observed hyperthermic potentiation produced by quercetin is consistent with the explanation that hyperthermic sensitization results from drug-induced lactate accumulation and intracellular acidifications with a resultant perturbation of the energy state of the cells. The increased effect under low pH conditions results from a low initial intracellular pH that is lowered further by drug effects carrying it to extreme values with cytotoxic consequences.

4. Polyamines

Polyamines are small polycationic peptides that are critically involved in many biologic processes. In particular, polyamines are involved in the stabilization of DNA and methylation of RNA. Under physiological conditions, intracellular concentrations of polyamines are in the millimolar range, in contrast to extracellular levels of polyamines several fold lower. Intracellular concentrations seemed to correlate with the cellular proliferative activity. In a series of cell culture studies, Gerner and Russell (1977) have initially demonstrated that when Chinese hamster ovary (CHO) cells in culture are exposed to 43 °C, they become leaky to the naturally occurring polycations, and correlates the loss and recovery of intracellular polyamine levels to DNA replication in heated cells. Subsequently, Ben-Hur *et al.* (1978) showed that the exogenous polyamines can enhance heat-induced cytotoxicity. This sensitization is dependent on the concentration and biophysical properties of the polyamines, such as charge and chain length (Gerner *et al.* 1980). Maximal sensitization develops when two terminal amino groups are separated by a distance of 10–11 Å. Thus, spermidine, 1,8-diamino octane and methylglyoxal bis-(guanylyhydrazone), compounds with terminal amino groups separated by 10–11 Å, all exhibit similar degrees of heat sensitization (Fig. 5).

Further studies with the inhibitor of polyamine synthesis have provided some important insights into the mechanism of the polyamine thermosensitization. Methylglyoxal bis-(guanylhydrazone) (MGBG) shares some structural similarity to the natural polyamines as well as to certain diamines, but MGBG is also a competitive inhibitor of the enzyme S-adenosyl methionine decarboxylase, which ultimately blocks the formation of spermidine from its precursor putrescine. Exposure of Chinese hamster ovary cells to MGBG was effective in inhibiting conversion of putrescine into spermidine and spermine. However, cells heated immediately following MGBG removal or

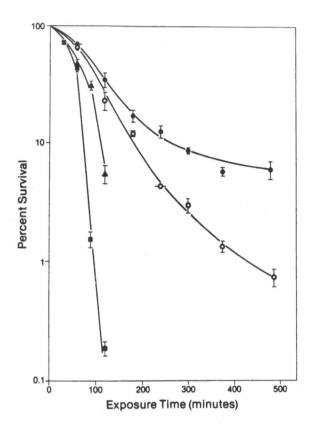

Figure 5: Per cent cell survival of Chinese hamster ovary cells exposed to 42 °C and various polyamines. ●: heat alone; ○: heat plus putrescine; ▲: heat plus spermidine; ■: heat plus spermine (from Gerner *et al.* 1980).

heated 30 hours after drug removal were sensitized to heat while cells heated at 12 through 24 hours after drug removal were moderately sensitized to heat (Gerner *et al.* 1983). Based on the findings that lowered intracellular polyamine levels did not correspond with subsequent survival responses to heat and exogenous spermine as low as 100 μM in the culture media dramatically sensitized cells expressing thermotolerance, Fuller and Gerner (1982) proposed that polyamines are molecular mediators of thermotolerance and cellular heat sensitivity. Their proposed model incorporates the differential interaction of polyamines with the outer and inner aspect of the plasma membranes. However, recent studies on the interaction between heat and polyamines provide alternative biochemical mechanisms whereby the synergism between heat and polyamines may be in part mediated by oxidation products of polyamines, since the sensitization effect is significantly reduced by simultaneous exposure of cells to polyamines and aminoguanidine, a putative inhibitor of serum diamine oxidase (Mondovi *et al.* 1981; Henle *et al.* 1986).

Since MGBG enhances the thermosensitivity of cells in culture and is undergoing phase III clinical trials as a chemotherapeutic agent, an *in vivo* tumor experiment was carried out with a murine tumor system (Kim *et al.* 1986). The tumor growth delay time of Meth-A fibrosarcoma in Balb/c mice was prolonged by an amount dependent on the drug concentration and exposure time of hyperthermic treatment (Fig. 6). A similar enhanced effect on the tumor growth delay was obtained with another murine tumor system (Miyakoshi *et al.* 1984). There was no disproportionately enhanced foot skin reaction observed, although another study showed enhanced skin reaction from heat and MGBG, when the drug was given 4 hours before the fractionated hyperthermia (Leith, 1982). Further *in vivo* tumor studies are needed to optimize the combined use of polyamine inhibitors and heat.

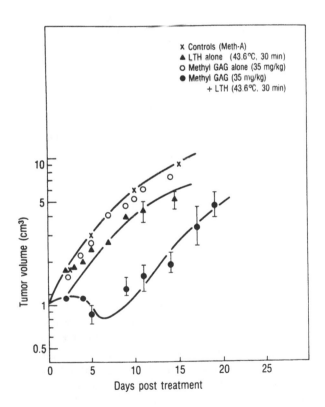

Figure 6: Effect of single dose administration of methylglyoxal bis-(guanylhydrazone) (Methyl GAG) and local tumor hyperthermia on the growth delay of methylcholanthrene induced fibrosarcoma (Meth-A) in Balb/c mice.
×: the growth rate of untreated, control tumor; ○: the growth rate of drug alone (35 mg/kg); ▲: heat alone (43.6 °C for 30 min.), water bath heating; ●: the combined effect of heat and drug. The Methyl GAG was injected i.p. immediately before heating (from Kim *et al.* 1986).

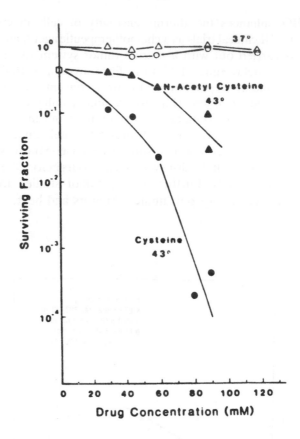

Figure 7: Cell surviving fraction of V79 cells treated for 1 hour at 43 °C in the presence of varying concentrations of cysteine and N-acetylcysteine. The 1 hour at 43 °C resulted in a surviving fraction of 0.45. Open symbols represent survival of cells treated with cysteine (O) or *N*-acetylcysteine (△) at 37 °C for 1 hour (from Mitchell and Russo 1983).

5.   *Thiol compounds*

Exposure of mammalian cells to exogenous thiol containing compounds such as cysteamine and cysteine potentiates heat-induced cell killing (Kapp and Hahn 1979). The drugs alone at 37 °C are not cytotoxic and there is only very limited drug–heat interaction up to 42 °C. However, a marked synergism of cysteamine is seen between 42 ° and 43 °C. In an attempt to understand the mechanism of heat potentiation induced by thiol containing compounds, Mitchell and Russo (1983) carried out similar studies with cysteine and *N*-acetylcysteine. They reasoned that the amino-thiol compound, cysteine, might be promoting a free radical reaction and the effect may be reduced by blocking the amine with an acetyl group, *N*-acetylcysteine. Figure 7 shows that the degree of sensitization is reduced if the amino group is blocked. Further studies suggested that activated oxygen species like superoxide and hydrogen peroxide, which are generated during autoxidation of thiols in the presence of oxygen, are involved in the mechanism of heat potentiation by thiol compounds (Issels *et al*. 1984).

In pursuing the biochemical mechanism of thermosensitization and thermotolerance, Mitchell and Russo (1983) evaluated the cellular-redox state during thermal stress by measuring and altering the cellular concentrations of glutathione (GSH), since GSH is intimately involved in maintenance of the cellular redox state and detoxification. Exposure of Chinese hamster V79 cells to continuous heating at 42.5 °C or acute exposure at 43 °C resulted in rapid elevations of cellular GSH to 120 per cent to 200 per cent of control values. Further, GSH depletion by two different modalities, diethylmalcate (DEM) and buthionine sulfoximine (BSO), which lowers GSH to values less than 5 per cent of controls, dramatically sensitized cells to heating at 42.5 °C. Interestingly, when DEM-treated GSH-depleted cells were heated at 42.5 °C, their GSH level rapidly increased. The magnitude of the thermosensitization by GSH depletion was diminished as the temperature was increased from 43 ° to 45 °C. A more recent study shows that the enhanced thermosensitivity by GSH depletion is dependent upon oxygen tension (Freeman *et al*. 1985). The effects of DEM and BSO also appear cell-line dependent. Using two cell lines, mouse fibroblast LM cells and Ehrlich ascites tumor cells, Konnings and Penninga (1985) were unable to influence thermosensitivity of cells depleted of GSII by DEM and BSO treatment, although the magnitude of GSII reduction was as low as 5 per cent of the control.

There exist several effective enzymatic systems to provide protection from oxygen-induced free radical species and peroxides that are normally produced at low levels as a result of oxygen metabolism. If these highly reactive compounds are not neutralized, they would be extremely deleterious to the cell. These enzymes are superoxide dismutase (SOD), glutathione peroxidase (GP), catalase and the cytochrome oxidase complex. Diethyldithiocarbamate (DDC) is a good copper-chelating agent and has been shown to inactivate many copper-containing enzymes including SOD and GP, and consequently DDC-treated cells would possibly incur a risk of oxidative injuries from oxygen radicals. Indeed, Lin *et al*. (1979) showed potentiation of heat killing at 43 °C in exponentially growing DON cells by prior treatment with DDC. Some of the potentiating effect may be, however, brought by inhibition of the repair of potentially lethal damage of heat (Evans *et al*. 1983).

C. HYPERTHERMIC PROTECTORS

Studies on the protective effects of hyperthermia by chemical means have been rather limited to a few compounds relative to hyperthermic sensitizers. The reason for the paucity of protective agents is not apparent, but in part may be related to limited understanding on the nature of 'target' molecule(s) or site of heat action and also in part to the extensive study on the thermotolerance, since heat by itself is the best known inducer of heat protection. None the less, several compounds have been shown to protect the cytotoxic effect of heat in various mammalian cells. As with hyperthermic sensitizers, results obtained with protective agents undoubtedly provide an important insight to understanding the mechanism of heat action on cells.

1. Deuterium oxide

Deuterium oxide (D_2O) is known to have many biological properties, one of which is the stabilization of macromolecules. Several investigators have shown that D_2O can protect mammalian cells against heat-induced cell killing (Ben-Hur and Riklis 1980; Azzam et al. 1982; Fisher, et al. 1982; Li et al. 1982; Raaphorst and Azzam 1982), although one study showed no effect (Lin et al. 1984).

Detailed quantitative studies of interaction of D_2O and hyperthermia in mammalian cells in culture were carried out by Hahn and his co-workers (Hahn et al. 1978; Fisher et al. 1982; Li et al. 1982). Survival curves as a function of temperature are similar in shape for the various D_2O concentrations, but the temperature threshold for cell killing is progressively shifted to higher temperatures in the presence of increasing concentrations of D_2O. The presence of 85 per cent deuterium oxide during a 3-minute exposure to elevated temperatures results in a displacement of 1.6 °C to 1.8 °C in the temperatures above 44 °C (Fig. 8). The protective effect of D_2O is seen only during heating. It is not enhanced by prior exposure to D_2O. An Arrhenius plot constructed from cells treated by D_2O and H_2O displays parallel lines,

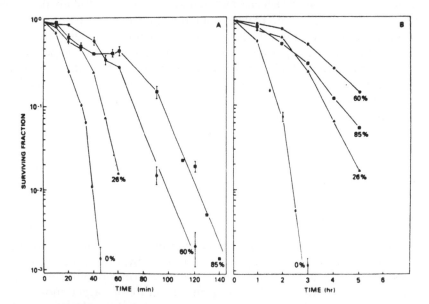

Figure 8: Effect of D_2O concentration on the survival of Chinese hamster cells (HA-1) exposed to 45 °C (panel A) or 43 °C (panel B). Exponentially growing CHO cells were exposed to graded concentrations of D_2O as a function of time. Note that the maximum heat protection was obtained with 60 per cent D_2O for cells exposed to 43 °C (from Fisher et al. 1982).

but shifted by 1.7 °C, suggesting that the cellular target in heat inactivation is not different but merely stabilized in the presence of D_2O. More recently, Lin *et al*. (1984) suggested that the heat protection by D_2O is mediated through the microtubular cytoskeleton, based on their study of the protective effect of D_2O on the microtubular structure and the similar protective effect from vinka alkaloids, whose main cytotoxic target has been regarded as on the microtubule.

2. *Glycerol and polyols*

As with deuterium oxide, glycerol and polyhydroxy compounds such as polyols and sugars protect catalytic and structural proteins against heat denaturation and whole cells against heat death (Back *et al*. 1979; Lin *et al*. 1979; Henle and Waters 1982). Using CHO cells and HeLa cells in culture, Henle and Waters have shown that exposure of cells to 1M-glycerol protected against thermal killing (Henle and Waters 1982). Exposure of cells to glycerol before or after heating at 45 °C did not protect against cell killing. Protection against heat killing requires glycerol to be present intracellularly during heating. Above glycerol concentrations of 100 mM, a protective effect of the drug increased in a concentration dependent manner. Glycerol also partially reversed the sensitizing effect of low pH and stepdown heating at 41.5 °C. However, long-term nutrient deprivation in the presence of 1M glycerol did not protect against cell killing at 41.5 °C but instead sensitized to heat. Glycerol protection against heat damage has been ascribed to increased hydrophobic interactions between solvent and solute macromolecules (Back *et al*. 1979). Such a physico-chemical reaction would tend to stabilize proteins, which implies that one of the heat-killing mechanisms is the inactivation of a heat-sensitive molecule. Glycerol has been also shown to reduce the heat-induced membrane permeability changes and to partially inhibit protein synthesis; both events can contribute to the protective effect of stabilizing cellular protein against heat killing (Lin *et al*. 1984). Lin *et al*. further postulated that microtubule-related protein is probably one of the proteins stabilized by glycerol, since glycerol can also reduce the toxicity of vincristine, an anti-tubular agent.

Based on the finding that polyhydroxy compounds and glycerol can protect against heat-induced cell killing, Henle (1981) postulated that naturally occurring protector molecules synthesized by thermotolerant cells are polyhydroxy compounds. Subsequent cell culture studies with polyols provided some supporting data, indicating that linear polyols constitute a family of heat-protecting compounds and that they may reduce heat killing only when they are present inside of cells (Henle *et al*. 1983). The non-linear tetrahydroxy alcohol pentaerythritol is not a naturally occurring compound and cannot protect against heat damage. In pursuing their polyol hypothesis for thermotolerance, Henle *et al*. (1984) presented further experimental data that various naturally occurring sugars can protect Chinese hamster ovary cells against heat killing. Unlike glycerol, heat protection by sugars was not immediate but required pre-incubation in the medium before heating. The degree of heat protection conferred by each sugar and its time dependence differed: galactose was the most effective heat protector and the protective effect was proportional to the sugar concentration in the medium up to 0.3M. Glucose and mannose were less effective heat protectors at 0.3M concentrations relative to galactose at the equimolar concentration. Under iso-osmotic conditions, however, heat protection by glucose appeared to be more prompt and apparent than with galactose. It is not apparent what specific intracellular metabolites of polyols and sugars would confer heat protection. The inhibition of glucose flux through the pentose monophosphate shunt pathway failed to suppress the development of thermotolerance, suggesting that intracellular polyols may not be the molecular mediator of thermotolerance since this pathway is considered to play a central role in the formation of polyols from aldo-sugars (Konnings and Penninga 1985).

3. Differentiating agents

Agents known to influence proliferation and differentiation of many transformed cell lines have been shown to afford protective effects on the thermosensitivity of cells (Kim et al. 1984). The treatment of cells in culture to cyclic adenosine 3':5'–monophosphate (cAMP), its derivatives, or agents that increase the intracellular cAMP has been shown to produce a multitude of morphological changes associated with cellular differentiation (Cho-Chung 1980; Johnson et al. 1971). Retinoic acid (RA) has been shown also to alter cellular morphology of various cell lines in culture, including cytoskeletal elements and external membrane components such as glycoproteins, fibronectin, or lamin (Kubilus et al. 1981; Tsao et al. 1982). Exposure of HeLa S–3 cells to dibutyryl cAMP (dbC-AMP), sodium butyrate and RA all afforded protective effects on the thermosensitivity of these cells. The kinetics of expression of thermal resistance induced by the agents varied with different agents (Fig. 9). Sodium butyrate (1mM) was only capable of inducing the thermal resistance during the time of heating (42 °C), while a minimum

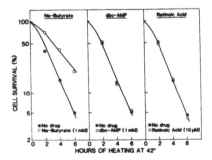 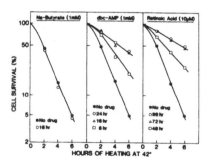

Figure 9: Effects of differentiating agents on thermosensitivity of cells. The drugs were present only during heating (Panel A). The effects of pre-treatment of cells to the drugs on the thermosensitivity (Panel B). The drugs were removed during heating. The hour indicates the duration of the drug treatment time at 37 °C prior to heat treatment at 42 °C (from Kim et al. 1984).

8-hour exposure to dbc-AMP (1mM) and 48-hour exposure to RA (10 mM) prior to heating at 42 °C were required to demonstrate the thermal resistance. Unlike glycerol and sodium butyrate, the presence of dbc-AMP and RA was not required during the heating. This difference in the kinetics of thermal resistance suggests that these agents may have a different mode of action for the induction of thermal resistance in HeLa cells. The protective effects of these agents appear to be cell line dependent, since exposure of neuroblastoma cells to cAMP produced a synergistic effect of heat (Rama and Prasad 1984). Other studies also failed to produce a full thermotolerant state by exposing cells to dbc-AMP, although the kinetics of cAMP increase after heating, suggesting a role in thermotolerance induction (Calderwood *et al.* 1985). More studies are needed with different cell lines to establish the relationship between the intracellular cAMP level and thermosensitivity of cells.

D. SUMMARY

A variety of chemical agents whose mode of action is well understood in cells at 37 °C have been employed to determine reasons why and how cells die when exposed to elevated temperatures. Findings from such studies have provided several important clues toward the identification of the principal mechanism of hyperthermia induced cytoxicity.

Several classes of drugs have been shown to be excellent hyperthermic sensitizers in cell culture studies. The class of compounds, known as membrane active agents, including local anesthetics, phenothiazines and short chained aliphatic alcohols, has been shown to potentiate hyperthermic cytotoxicity by lowering temperature thresholds. The synergistic interaction of membrane active agents with heat in a variety of cellular systems underscores the importance of the membrane as a critical target in the enhancement of heat-induced cell killing. The class of compounds known as energy depleters also includes excellent heat sensitizers. Most inhibitors of glycolysis (5-thio-D-glucose, pentalenolactone) selectively enhance the thermosensitivity of hypoxic cells, since cells under hypoxia are solely dependent on their source of energy via anaerobic glycolysis. The thermosensitivity of oxic cells is not influenced by the drugs. There are, however, other glycolytic inhibitors (lonidamine, gossypol) whose sensitizing effect by heat is seen in both oxic and hypoxic cells. The hyperthermic sensitizing effect of these inhibitors is interestingly most pronounced under the low pH of culture medium. The major hyperthermic sensitization is usually obtained with inhibitors of oxidative phosphorylation only in the absence of glucose in the media, indicating that lowering the cellular energy level to the critical level renders cells extremely heat sensitive. The use of lactate transport inhibitors as a hyperthermic sensitizer supports the concept that the energy equilibrium of the cell is an important factor in determining the thermosensitivity of cells, since hyperthermic sensitization results from drug-induced lactate accumulation and intracellular acidifications with a resultant perturbation of the energy state of the cells. Exogenous polyamines can enhance heat-induced cytotoxicity and the sensitization is thought to be a differential interaction of polyamines with a specific segment of the outer aspect of the plasma membrane. Lastly, exogenous thiol containing compounds (cysteamine, cysteine) can potentiate heat-induced cytotoxicity. The postulated mechanism of heat potentiation by thiol compounds is the enhanced production of activated oxygen species during autoxidation of thiols in the presence of oxygen. The study indicates the importance of the cellular redox state during the thermal stress.

Hyperthermic protectors are limited in numbers to a few class of agents. Among them, deuterium oxide and glycerol, both having the property of stabilizing the macromolecules including protein, can protect mammalian cells against heat-induced cell killing. The protective effect of the agents is seen only during heating and there is an upward shift of the temperature threshold. Other classes of drugs that can protect against the heat damage are polyols and naturally occurring sugars. The suggestion as to whether the intracellular polyols are a molecular mediator of thermotolerance is not well substantiated. Finally, several agents known to influence proliferation and differentiation of many transformed cell lines have been shown to afford protective effects on the thermosensitivity of cells. The biochemical and cellular basis for the protective effect of c-AMP and retinoic acid is not well understood, although changes in the cytoskeleton of microtubule have been implicated.

REFERENCES

Abou-Donia, M.B. and Diechert, J.W., 1974. Gossypol: uncoupling of respiratory chain and oxidative phosphorylation. *Life Sci.* **14**, 1955–1963.

Azzam, E.I., George I., and Raaphorst, G.P., 1982. Alteration in thermal sensitivity of Chinese hamster cells by D_2O treatment. *Radiat. Res.* **90**, 644–648.

Back, J.F., Oakenfull, D., and Smith, M.B., 1979. Increased thermal stability of proteins in the presence of sugars and polyols. *Biochemistry* **18**, 5191–5196.

Belt, J.A., Thomas, J.A., Buchsbaum, R.N., and Racker, E., 1979. Inhibition of lactate transport and glycolysis in Ehrlich ascites tumor cells by bioflavonoids. *Biochemistry* **18**, 3506–3511.

Bernal, S.D., Shapiro, H.M., and Chen, L.B., 1982. Monitoring the effects of anti-cancer drugs on L1210 cells by a mitochondrial probe, rhodamine 123. *Int. J. Cancer* **30**, 219–224.

Ben-Hur, E., Prager, A., and Riklis, E., 1978. Enhancement of thermal killing by polyamines. I. Survival of Chinese hamster cells. *Int. J. Cancer* **22**, 602–606.

Ben-Hur, E., and Riklis, E., 1978. Enhancement of thermal killing by polyamines. II. Uptake and metabolism of exogenous polyamines in hyperthermic Chinese hamster cells. *Int. J. Cancer* **22**, 607–610.

Ben-Hur, E. and Riklis, E., 1979. Enhancement of thermal killing by polyamines. III. Synergism between spermine and gamma radiation in hyperthermic Chinese hamster cells. *Radiat. Res.* **78**, 321–328.

Ben-Hur, E. and Riklis, E., 1980. Deuterium oxide enhancement of Chinese hamster cell response to X-irradiation. *Radiat. Res.* **81**, 224–235.

Calderwood, S.K., Stevenson, M.A. and Hahn, G.M., 1985. Cyclic AMP and the heat shock response in Chinese hamster ovary cells. *Biochem. Biophys. Res. Commun.* **216**, 911–916.

Chen, L.B., Summerhayes, I.C., Johnson, L.V., Walsh, M.L., Bernal, S.D. and Lampidis, T., 1982. Probing mitochondria in living cells with rhodamine 123. *Cold Spring Harbor Symposia on Quantitative Biology* **46**, 141–155.

Chen, T.T. and Heidelberger, C., 1969. Quantitative studies on the malignant transformation of mouse prostate cells by carcinogenic hydrocarbons *in vitro*. *Int. J. Cancer* **4**,166–178.

Cho-Chung, Y.S., 1980. Cyclic AMP and its receptor protein in tumor growth regulation *in vivo. J. Cyclic Nucleotide Res.* **6**, 163–173.

Coss, R.A. and Dewey, W.C., 1982. Heat sensitization of G1 and S-phase cells by procaine hydrochloride. *Radiat. Res.* **92**, 615–617.

Darzynkiewicz, Z., Traganaos, F., Staiano-Coico, L., Kapuscinski, J., and Melamed, M.R., 1982. Interactions of rhodamine 123 with living cells studied by flow cytometry. *Cancer Res.* **42**, 799–806.

Dennis, W.H. and Yatvin, M.B., 1981. Correlation of hyperthermic sensitivity and membrane microviscosity in E. Coli K 1060. *Int. J. Radiat. Biol.* **39**, 265–271.

Dewey, W.C., Sapareto, S.A. and Gerweck, L.E., 1977. Cellular response to combination of hyperthermia and radiation. *Radiation* **123**, 463–474.

Dewey, W.C., Freeman, M.L., Raaphorst, G.P., Clark, E.P., Wong, R.S.L., Highfield, E.P., Spiro, I.J., Tomasovic, S.P., Denamn, E.L. and Coss, R.A., 1980. Cellular biology of hyperthermia and radiation. In *Radiation Biology in Cancer Research*, (Eds.) Meyn, R.E. and Withers, H.R. Raven Press, New York, pp. 589–621.

Dubinsky, W.P. and Racker, E., 1978. The mechanism of lactate transport in human erythrocytes. *J. Membr. Biol.* **44**, 25–30.

Evans, R.G., Nielsen, J., Engel, C. and Wheatley, C., 1983. Enhancement of heat sensitivity and modification of repair of potentially lethal heat damage in plateau-phase cultures of mammalian cells by diethylthiocarbamate. *Radiat. Res.* **93**, 319–325.

Fisher, G.A., Li, G.C., and Hahn, G.M., 1982. Modification of the thermal response by D_2O. 1. Cell survival and the temperature shift. *Radiat. Res.* **92**, 530–540.

Floridi, A., Paggi, M.G., Marcante, M.L., Silvestrini, B., Caputo, A. and Martino, C.D., 1981. Lonidamine, a selective inhibitor of aerobic glycolysis of murine tumor cells. *J. Natl. Cancer Inst.* **66**, 497–499.

Freeman, M.L., Malcolm, A.W., and Meredith, M.J., 1985. Role of glutathione in cell survival after hyperthermic treatment of Chinese hamster ovary cells. *Cancer Res.* **45**, 6308–6313.

Fuller, D.J.M., and Gerner, E.W., 1982. Polyamines: a dual role in the modulation of cellular sensitivity to heat. *Radiat. Res.* **92**, 439–444.

Gerner, E.W. and Russell, D.H., 1977. The relationship between polyamine accumulation and DNA replication kinetics in synchromized CHO cells after heat shock. *Cancer Res.* **37**, 482–489.

Gerner, E.W., Cress, A.E., Stickney, D.G., Holmes, D.K., and Culver, P.S., 1980. Factors regulating membrane permeability after thermal resistance. *Ann. N.Y. Acad. Sci.* **335**, 215–230.

Gerner, E.W., Holmes, D.K., Stickney, D.G., Noterman, J.A., and Fuller, D.J.M., 1980. Enhancement of hyperthermia induced cytotoxicity by polyamines. *Cancer Res.* **40**, 432–438.

Gerner, E.W., Stickney, D.G., Herman, T.S. and Fuller D.J.M., 1983. Polyamines and polyamine biosynthesis in cells exposed to hyperthermia. *Radiat. Res.* **93**, 340–352.

Gerweck, L.E., 1977. Modification of cell lethality of elevated temperatures: The pH effect. *Radiat. Res.* **70**, 224–235.

Gerweck, L.E., Dahlberg, W.K., Epstein, L.F. and Shimm, D.S., 1984. Influence of nutrient and energy deprivation on cellular response to single and fractionated heat treatments. *Radiat. Res.* **99**, 573–581.

Giovanella, B.C., Stehlin, J.S. and Morgan, A.C., 1976. Selective lethal effect of supranormal temperatures on human neoplastic cells. *Cancer Res.* **36**, 3944–3950.

Hahn, G.M., 1974. Metabolic aspects of the role of hyperthermia in mammalian cell inactivation and their possible relevance to cancer treatment. *Cancer Res.* **34**, 3117–3123.

Hahn, G.M., Li, G.C. and Shiu, E.C., 1977. Interaction of amphotericin B and 43 °C hyperthermia. *Cancer Res.* **37**, 761–764.

Hahn, G.M., Steinberg, D., and Fisher, G., 1978. Protection by deuterium oxide against hyperthermic inactivation of Chinese hamster cells. *Radiat. Res.* **74**, 476–477.

Hahn, G.M., 1980. Comparison of malignant potential of 10 T 1/2 cells and transformants with their survival responses to hyperthermia and to amphotericin B. *Cancer Res.* **40**, 3763–3767.

Hahn, G.M., 1982. *Hyperthermia and Cancer*. Plenum Press, New York.

Hahn, G.M., and Li, G.C., 1982. The interactions of hyperthermia and drugs: Treatments and probes. *J. Natl. Cancer Inst. Monogr.* 317–324.

Haveman, J. and Hahn, G.M., 1981. The role of energy in hyperthermia-induced mammalian cell inactivation: A study of the effects of glucose starvation and an uncoupler of oxidative phosphorylation. *J. Cell Physiol.* **107**, 237–241.

Henle, K.J., 1981. Interaction of mono- and polyhydroxy alcohols with hyperthermia in CHO cells. *Radiat. Res.* **88**, 392–402.

Henle, K.J., Nagle, W.A., Moss, A.J. Jr., and Herman, T.S., 1982. Polyhydroxy compounds and thermotolerance: a proposed concatenation. *Radiat. Res.*, **92**, 445–451.

Henle, K.J., Peck, J.W., and Higashikubo, R., 1983. Protection against heat-induced cell killing by polyols *in vitro*. *Cancer Res.* **43**, 1624–1627.

Henle, K.J., Monson, T.P., Moss, A.J. and Nagle, W.A., 1984. Protection against thermal cell death in Chinese hamster ovary cells by glucose, galactose or mannose. *Cancer Res.* **44**, 5499–5504.

Henle, K.J., Moss, A.J. and Nagle, W.A., 1986. Mechanism of spermidine cytotoxicity at 37 °C and 43 °C in Chinese hamster ovary cells. *Cancer Res.* **46**, 175–182.

Hong, S.K., Haspel, H.C., Sonenberg, M. and Goldinger, M., 1983. Effect of gossypol on PAH transport in the rabbit kidney slice. *Toxicol. Appl. Pharmacol.* **71**, 430–435.

Johnson, J.H., Belt, J.A., Dubinsky, W.P., Zimniack, A., and Racker, E., 1980. Inhibition of lactate transport in Ehrlich ascites tumor cells and human erythrocytes by a synthetic anhydrides of lactic acid. *Biochemistry* **19**, 3836–3840.

Johnson, G.S., Friedman, R.M., and Pastan, I., 1971. Restoration of several morphological characteristics of normal fibroblasts in sarcoma cells treated with adenosine 3′–5′–cyclic monophosphate and its derivatives. *Proc. Natl. Acad. Sci. U.S.A.* **68**, 425–429.

Kalla, N.K., 1982. Gossypol – the male antifertility agent. *IRCS. Med. Sci.* **10**, 766–769.

Kapp, D.S. and Hahn, G.M., 1979. Thermosensitization by sulfhydryl compounds of exponentially growing Chinese hamster cells. *Cancer Res.* **39**, 4630–4635.

Kim, J.H., Kim, S.H., Hahn, E.W. and Song, C.W., 1978. 5-Thio-D-glucose selectively potentiates hyperthermic killing of hypoxic tumor cells. *Science* **200**, 206–207.

Kim, J.H., Kim, S.H. and Hahn, E.W., 1978. Killing of glucose deprived hypoxic cells with moderate hyperthermia. *Radiat. Res.* **75**, 448–451.

Kim, J.H., Alfieri, A.A., Kim, S.H. and Young, C.W., 1984. Lonidamine, a hyperthermic sensitizer of HeLa cells in culture and of the Meth-A tumor *in vivo*. *Oncology* **41**, 30–35.

Kim, J.H., Kim, S.H., Alfieri, A.A., and Young, C.W., 1984. Quercetin, an inhibitor of lactate transport is a hyperthermic sensitizer of HeLa cells. *Cancer Res.* **44**, 102–106.

Kim, J.H., Kim, S.H. and Alfieri, A.A., 1985. Interaction of rhodamine 123 and hyperthermia in HeLa cells in culture. *Int. J. Hyperthermia* **1**, 247–253.

Kim, S.H., Kim, J.H. and Hahn, E.W., 1978. Selective potentiation of hyperthermic killing of hypoxic cells by 5-Thio-D-glucose. *Cancer Res.* **38**, 2935–2938.

Kim, S.H., Kim, J.H., Hahn, E.W. and Ensign, N.A., 1980. Selective killing of glucose and oxygen-deprived HeLa cells by hyperthermia. *Cancer Res.* **40**, 3459–3462.

Kim, S.H., He, S.Q., and Kim, J.H., 1984. Modification of thermosensitivity of HeLa cells by sodium butyrate, dibutyryl cyclic AMP and retinoic acid. *Cancer Res.* **44**, 697–702.

Kim, S.H., Kim, J.H., Alfieri, A.A. and Young, C.W., 1985. Gossypol, a hyperthermic sensitizer of HeLa cells. *Cancer Res.* **45**, 6338–6340.

Konnings, A.W.T. and Penninga, P., 1985. On the importance of the level of gluthathione and the activity of the pentose phosphate pathway in heat sensitivity and thermotolerance. *Int. J. Radiat. Biol.* **48**, 409–422.

Kubilus, J., Rand, R. and Baden, H.P., 1981. Effects of retinoic acid and other retinoids on the growth and differentiation of 3T3 supported human keratinocytes. *In Vitro*, **17**, 786–795.

Lampidis, T.J., Bernal, S.D., Summerhayes, I.C., and Chen, L.B., 1983. Selective toxicity of rhodamine 123 in carcinoma cells *in vitro*. *Cancer Res.* **43**, 716–720.

Laval, F., and Michel, S., 1982. Enhancement of hyperthermia-induced cytotoxicity upon ATP deprivation. *Cancer Lett.* **15**, 61–65.

Leith, J.T., 1982. Effects of methylgloxal bis (guanylhydrazone) on skin reactions in the mouse after fractionated hyperthermic exposures. *Radiat. Res.* **90**, 586–594.

Lepock, J.R., 1982. Involvement of membranes in cellular response to hyperthermia. *Radiat. Res.* **92**, 433–438.

Li, G.C. and Hahn, G.M., 1978. Ethanol-induced tolerance to heat and to adriamycin. *Nature* **274**, 699–701.

Li, G.C., Shiu, E.C. and Hahn, G.M., 1980. Similarities in cellular inactivation by hyperthermia or by ethanol. *Radiat. Res.* **82**, 257–268.

Li, G.C., Fisher, G.A., and Hahn, G.M., 1982. Modification of the thermal response by D_2O II. Thermotolerance and the specific inhibition of development. *Radiat. Res.* **92**, 541–551.

Lin, P.S., Kwock, L. and Butterfield, C.E., 1979. Diethyldithiocarbamate enhancement of radiation and hyperthermic effects on Chinese hamster cells *in vitro*. *Radiat. Res.* **77**, 501–511.

Lin, P.S., Hefter, K. and Ho, K.C., 1984. Modification of membrane function, protein synthesis, and heat killing effect in cultured Chinese hamster cells by glycerol and D$_2$O. *Cancer Res.* **44**, 5776–5784.

Mondovi, B., Guerieri, P., Costa, M.T. and Sabatini, S., 1981. Amine oxidase inhibitors and biogenic amines metabolism. *Adv. Polyamine Res.* **3**, 75–84.

Mitchell, J.B., and Russo, A., 1983. Thiols, thiol depletion, and thermosensitivity. *Radiat. Res.* **95**, 471–485.

Miyakoshi, J., Hiraoka, M., Oda, W., Takahasi, M., Abe, M. and Inagaki, C., 1984. Effects of methylglyoxal bis (guanylhydrazone) on tumor and skin responses to hyperthermia in mice. *Int. J. Radiat. Biol.* **46**, 287–291.

Nagle, W.A., Moss, A.J., and Baker, M.L., 1982. Increased lethality from hyperthermia at 42 °C for hypoxic Chinese hamster cells heated under conditions of energy deperivation. *Natl. Cancer Inst. Monogr.* **61**, 107–110.

Nagle, W.A., Moss, A.J. and Henle, K.J., 1985. Sensitization of cultured Chinese hamster cells to 42 °C hyperthermia by pentalenolactone, an inhibitor of glycolytic ATP synthesis. *Int. J. Radiat. Biol.* **48**, 821–835.

Raaphorst, G.P. and Azzam, E.I., 1982. The effect of deuterium oxide on thermal sensitivity and thermal tolerance in cultured Chinese hamster V79 cells. *J. Therm. Biol.* **7**, 147–154.

Rama, B.N. and Prasad, K.N., 1984. Effect of hyperthermia in combination with Vitamin E and cyclic AMP on neuroblastoma cells in culture. *Life Sci.* **34**, 2089–97.

Robins, H.I., Dennis, W.H. and Vorpahl, J.W., 1982. Systemic lidocaine enhancement of tumor regression after hyperthermia. *Natl. Cancer Inst. Monogr.* **61**, 243–245.

Seeman, P., 1972. The membrane actions of anesthetics and tranquilizers. *Pharmacol. Rev.* **24**, 583–655.

Sheetz, M.P. and Singer, S.J., 1974. Biological membranes as bilayer couples. A molecular mechanism of drug-erythrocyte interactions. *Proc. Natl. Acad. Sci. U.S.A.* **71**, 4457–4461.

Silva, M.T., Sousa, J.C.F., Polonia, J.J. and Macedo, P.M., 1979. Effects of local anesthetics on bacterial cells. *J. Bacteriol.* **137**, 461–468.

Silvestrini, B., Hahn, G.M. and DeMartino, C., 1983. Effects of lonidamine alone or combined with hyperthermia in some experimental cell and tumour cells. *Br. J. Cancer* **47**, 221–231.

Song, C.W., Clement, J.J., and Levitt, S.H., 1977. Cytotoxic and radiosensitizing effects of 5-thio-D-glucose on hypoxic cells. *Radiology* **123**, 201–205.

Tsao, D., Morita, A., Bella, A., Luu, P. and Kim, Y.S., 1982. Differential effects of sodium butyrate, dimethyl sulfoxide, and retinoic acid on membrane-associated antigen, enzymes, and glucoproteins of human rectal adenocarcinoma cells. *Cancer Res.* **42**, 1052–1058.

Wallach, D., 1977. Basic membranes in tumor thermotherapy. *J. Molecular Med.* **2**, 381–403.

Yatvin, M.B., 1977. The influence of membrane lipid composition and procaine on hyperthermic death of cells. *Int. J. Radiat. Biol.* **32**, 513–521.

Yatvin, M.B., Clifton, K.H. and Dennis, W.H., 1979. Hyperthermia and local anesthetics: potentiation of survival of tumor-bearing mice. *Science* **205**, 195–196.

Yau, T.M., 1979. Procaine-mediated modification of membranes and of the response to X-irradiation and hyperthermia in mammalian cells. *Radiat. Res.* **80**, 423–541.

Yau, T.M. and Kim, S.C., 1980. Local anesthetics as hypoxic radiosensitizers, oxic radioprotectors and potentiators of hyperthermic killing in mammalian cells. *Br. J. Radiol.* **53**, 687–692.

Hyperthermia and Oncology, Vol. 1, pp. 121–159 (1988)
Urano and Douple (Eds)

Chapter 5

The response of normal tissues to hyperthermia

MARILYN P. LAW
*MRC Cyclotron Unit, Hammersmith Hospital, Du Cane Road, London W1_*
0HS, UK

A. INTRODUCTION

During the last ten years there has been renewed interest in using hyperthermia
to treat cancer, stimulated by experimental studies which suggest that heat
damages tumors more than normal tissues. In particular, solid tumors may
have an inadequate vascular supply so that many tumor cells are deprived
of nutrients, including oxygen, and thus respire anaerobically reducing tumor
pH. Such an environment increases thermal sensitivity of the cells, as
discussed in Chapter 6. In addition, an impaired blood supply results in a
reduced cooling ability so that tumors may reach higher temperatures than
normal tissues in localized heating fields. Other results indicate that heat
could be used to complement other more conventional therapy; cells in the
S-phase of the mitotic cycle are resistant to X-rays but sensitive to heat and
heat increases cellular sensitivity to chemotherapeutic agents.

Comparative studies of normal and neoplastic tissues *in vivo* are required
to design treatment regimes which optimize the use of hyperthermia in cancer
therapy. As with other anti-cancer agents the limiting factor in the clinical
use of hyperthermia will be the response of normal tissues included in the
treatment field. A range of heating techniques, based on the use of electromagnetic radiations or ultrasound, are available for clinical application
(Strohbehn and Douple 1984; Hand 1986). With their increasing use, information about the response of normal tissues in man is becoming available.
The majority of quantitative studies of tissue response, however, has been
in rodents heated by immersion in hot water. This review, therefore, concen-
trates on experimental studies of normal tissue response to hyperthermia,
although brief comparisons with tumor response will be made where appropriate. Detailed discussion will be restricted to responses to heat given alone
since combination with other modalities, such as radiotherapy or
chemotherapy, will be discussed in Volume 2 of the present series.

B.   EXPRESSION OF LOCAL THERMAL INJURY

When local hyperthermia is applied to treat superficial tumors the cardinal signs of the inflammatory response, which include redness and swelling, are observed at very early times after heating. This contrasts with the delay in the onset of the response to the ionizing radiations usually used to treat local malignancies. Temporal differences between the responses to heat or X-rays are illustrated for the skin of the mouse ear in Fig. 1. Mild hyperthermia causes transient erythema whereas more severe heating causes deep reddening, swelling and possibly necrosis during the first few days after treatment. If there is recovery, no further changes are observed after about 10 days. Irradiation, however, does not cause erythema until about 10 days after exposure and marked swelling is not observed. Depending on dose, radiation induces desquamation which reaches a maximum at 25 to 30 days and heals within 6 to 8 weeks. Figure 1 also shows that moderate hyperthermia, which alone causes little or no visible effect, may enhance radiation injury. The response to the combined treatment is qualitatively similar to that to radiation alone, there being erythema and desquamation followed by healing. With heat, however, a lower dose of radiation is required to cause a given reaction.

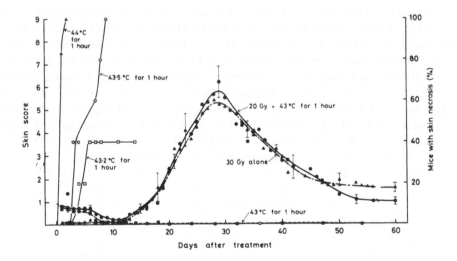

Figure 1: The response of the mouse ear to hyperthermia. Open symbols and right-hand axis: The response to heat alone is expressed as the percentage of ears showing necrosis. Closed symbols and left-hand axis: The response to X-rays given alone or to X-rays followed immediately by heat is assessed using an arbitrary numerical score. Scores of 1 and 2 indicate degrees of erythema and scores of 3 to 9 indicate increasing areas of moist desquamation. (Reproduced from Law *et al*. 1978.)

Similar differences in the time course of the response to heat or X-irradiation have been demonstrated in a number of rodent tissues. These include skin at various sites (Law *et al*. 1978; Jansen 1980; Miyakoshi *et al*. 1983; Hume and Myers 1984; Wondergem and Havemen 1984), cartilage (Morris *et al*. 1977; Myers 1983), testis (Hand *et al*. 1979), small intestine (Hume *et al*.

1979a), kidney (Elkon *et al.* 1980) and spinal cord (Sminia *et al.* 1985).

The delay in the onset of the radiation response is generally accepted to be related to the turnover of the target cells, cell death occurring only when irradiated cells attempt division. Consequently, with a knowledge of cell kinetics and of proliferative organization of the tissue, it is often possible to predict the time of expression of radiation injury which in the case of skin takes 2 to 3 weeks (Denekamp 1975). The earlier expression of thermal damage suggests that hyperthermia causes intermitotic cell death. Studies of cells in culture indicate that the time at which cells die after hyperthermia depends on the phase of the cell cycle (Cross and Dewey 1983). Whereas the majority of cells heated in S lyse after entering mitosis, the majority of cells heated in G_1 lyse before undergoing mitosis. This is consistent with heated S-phase cells dying at mitosis as a consequence of DNA damage and heated G_1 cells dying from a non-DNA related mechanism (Dewey *et al.* 1971; 1978).

Studies on small intestinal mucosa, *in vivo*, indicate that heat results not only in intermitotic death of the proliferative cells in the crypts but also has a directly cytocidal effect on the postmitotic functional epithelial cells of the villi (Hume *et al.* 1979a; Carr *et al.* 1982; van Beuningen *et al.* 1983). To further investigate this possibility, Hume and colleagues (1983) made use of the fact that after intraperitoneal injection of tritiated thymidine the only cells of the intestinal mucosa which are initially labeled are the proliferating cells of the crypts. Subsequently, the labeled non-proliferating progeny of these cells move on to the villi while the dividing cells in the crypts continue to dilute the label by consecutive cycles of DNA synthesis. By applying heat at various times after injection of ^{3}H-thymidine and measuring the subsequent loss of radioactivity it was possible to determine the relative thermal susceptibilities of the crypt and villus epithelial cell compartments. The time of expression of thermal injury was found to be similar in both compartments but the non-proliferating functional epithelial cells covering the villi were more susceptible to thermal injury than the crypt cells.

These results do not necessarily imply an intrinsic difference in heat sensitivity; stromal factors must also be considered. At times of maximal villous injury the villous stroma is edematous and haphazardly arranged with dilated capillaries and lacteals and increased cellularity due to infiltration of eosinophils, granulocytes and lymphocytes (Carr *et al.* 1986). Scanning electron microscopy has demonstrated changes in villous morphology as early as 5 minutes after heating, with maximal damage at 2 hours (Carr *et al.* 1986). Heating for 1 hour in the range 37 °C to 41 °C caused minimal damage (Fig. 2a). As the temperature was increased to 42 °C for 1 hour there was an increasing number of vertically collapsed villi (Fig. 2b). Irreversibly damaged and necrotic extruded epithelial cells can be seen at the villous tips. After 43 °C for 30 minutes villi were predominantly rudimentary in shape and occasionally absent altogether (Kamel *et al.* 1985).

Vascular injury has been implicated as a major factor in the development of tissue damage after hyperthermia in the clinical range (Song 1984). A

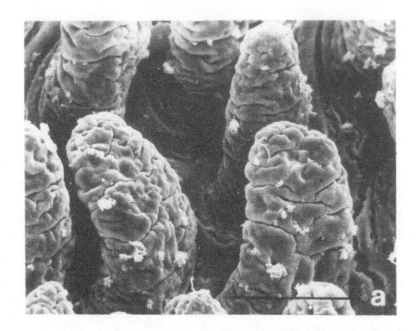

Figure 2: Scanning electron micrographs of small intestinal villi of mouse. Anesthetized mice were heated, up to the xiphisternum, in a water bath and specimens of jejunum sampled 2 hours after heating. a. 37.5 °C for 1 hour. Apart from the prominent creases, villi look normal. b. 41.5 °C for 1 hour. Villi are severely damaged with prominent extrusion zones at their tops. Some of the villi show sloughing. Bar = 100 μm. (Reproduced from Kamel *et al*. 1985.)

correlation between vascular lesions and gross tissue response in the intestine of mouse supports this view (Fig. 3). In the jejunum, capillaries and then progressively larger vessels are lost during a period of 5 days after heating (Falk 1983). A gradient of sensitivity occurs from the inner layer of the jejunum to the outer, the threshold heating temperature required to produce vascular damage being lowest at the lumen. Collapse of the villi (Kamel *et al.* 1985) shows a thermal dose dependence similar to that for the loss of the villous vasculature (Falk 1983), the threshold being about 40 °C for 1 hour. In contrast, crypt survival (Hume *et al.* 1979a) is less sensitive, the threshold being about 42 °C for 1 hour, comparable with that for vessels in the submucosa.

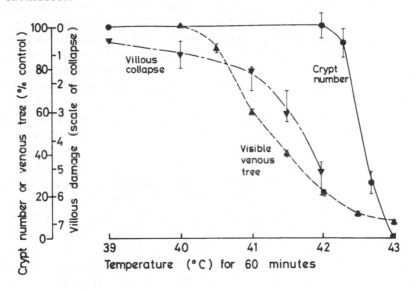

Figure 3: The response of small intestine of mouse to hyperthermia. The number of surviving crypts (●) or the length of venous tree, visualized by a benzidine staining technique, (▲) are expressed as a percentage of appropriate control values (data from Hume *et al.* 1979, Falk 1983). Villous collapse (▼) is assessed using an arbitrary scoring system (data from Kamel *et al.* 1985). (From S.P. Hume, unpublished.)

The importance of the microcirculation in the expression of thermal damage has also been demonstrated in tumors growing in transparent chambers implanted in the skin of the rat or cheek pouch of the hamster; hyperthermia caused failure of the microcirculation followed by tumor necrosis only in the regions showing circulatory failure (Reinhold *et al.* 1978; Eddy 1980). There is also a correlation between vascular occlusion in mouse tumors and decreased tumor cell survival assayed *in vitro* after heating *in vivo* (Kang *et al.* 1980).

It is likely that a number of factors causes hyperthermia-induced changes in the microcirculation. Alterations in micro-haemodynamics such as increased vascular resistance, decreased arterio-venous pressure gradients, arteriolar vasoconstriction and edema formation could all contribute to stasis of the blood as discussed by Reinhold and Endrich in a recent review (1986). In addition, aggregation of erythrocytes and thrombosis in capillaries are frequent findings (Dewhirst *et al.* 1984; Endrich and Hammersen 1984; Endrich *et al.* 1984). The extent to which these changes are related to heat-induced damage to the endothelial lining of vessels is unknown. Abnormalities in endothelial cells and leukocyte sticking have been observed at early times after heating (Song 1978; Endrich *et al.* 1979; Eddy 1980; Jain and Ward-Hartley 1984). The role of degenerative changes in endothelial cells, however, is equivocal. Rhee and Song (1984) found endothelial cells *in vitro* to be thermoresistant whereas Fajardo *et al.* (1985) found them thermosensitive. In addition, Badylak *et al.* (1985) report morphological changes in endothelial cells of the Walker carcinoma after heat doses which do not cause any measurable changes in tumor blood flow (Gullino *et al.* 1978; Gullino 1980; Song *et al* 1980a). Recent experiments, however, show that hyperthermia does inhibit the growth of blood vessels in mice, as measured by the inability of heated subcutaneous vasculature to support the growth of an implanted tumor, i.e., the tumor bed effect (Hill *et al.* 1985; Wondergem and Haveman 1985).

Although there are many *in vitro* studies of molecular and cellular mechanisms underlying the biological response to hyperthermia as reviewed elsewhere (Leeper 1984; Streffer 1985) there are few comparable studies *in vivo*. The work of Overgaard and Overgaard (1972, 1976a) implicated lysosomes in hyperthermic injury; both the number of lysosomes and lysosomal enzymatic activity were increased in a mouse mammary carcinoma following local heating. In contrast, quantitative biochemical techniques, which may themselves disrupt the relatively fragile lysosomes, failed to demonstrate activation of the lysosomal enzymes acid phosphatase or β-glucuronidase in mouse spleen and liver after whole-body hyperthermia (Tamulevicius and Streffer 1983). Histochemical staining methods which are non-destructive, however, have shown increased activities of acid phosphatase or β-glucuronidase in mouse spleen or rat cartilage (Hume *et al.* 1978a; Myers *et al.* 1981). Temperatures between 41 °C and 42.3 °C for 90 minutes increased enzyme activity, whereas higher temperatures increased the permeability of the lysosomal membrane, thus releasing hydrolytic enzymes into the cytoplasm (Hume *et al.* 1978a). Whether the release of lysosomal enzymes is the cause or a consequence of thermal injury in tissues is not established, but studies using a membrane labilizer, retinol, suggest that lysosomal membrane injury is unlikely to be a primary event in hyperthermal cell killing; administration of retinol increased the thermal response of lysosomes but had no effect on thermal damage at a gross tissue level (Rogers *et al.* 1983).

Detailed tissue studies at the level of the electron microscope, which may give further insight into the mechanisms of hyperthermal damage in tissues, are limited but ultrastructural changes have been reported at early times after hyperthermia. For example, in liver these include autophagic vacuoles, dilatation of Golgi cisternae, distension of rough endoplasmic reticulum and deposits in mitochondria (Wills *et al.* 1976).

The pathological effects of hyperthermia in various normal tissues have been reviewed by Fajardo (1984). In general, they are similar to, but less severe than, those observed after thermal burns. Depending on the heat treatment, edema, focal hemorrhage, necrosis and granulocytic infiltration occur during the first 24 hours. Later changes include loss of blood vessels, necrosis, monocyte infiltration and fibrosis. The morphological data, however, are incomplete and the ultimate result of focal hyperthermia is unknown for most tissues. Observation of gross response suggests that once the acute response to heat has healed there are no further changes (Law *et al.* 1977; Morris and Field 1985; Hume and Myers 1985), but sequential histological studies are required to determine whether early thermal damage results in permanent or even progressive microscopic lesions months or years after treatment.

C. TIME–TEMPERATURE RELATIONSHIPS

The response of mammalian cells and tissues to hyperthermia depends on both the temperature and duration of heating. This is illustrated for the mouse jejunum in Fig. 4. Crypt survival curves obtained by plotting crypt number as a function of heating time for a given temperature are similar to cell survival curves obtained after heating *in vitro* (see Chapter 3). There is an initial 'shoulder' region followed by an exponential region whose slope increases (i.e., thermal D_0 decreases) as temperature is increased. For longer heating times at temperatures of 42.5 °C and below, there is a plateau in the crypt susceptibility to thermal damage as heating time is increased, indicating the development of thermotolerance similar to that observed in cells *in vitro* after prolonged exposure to lower temperatures (40 °C to 42.5 °C). A transient resistance characterized by a plateau in the thermal survival curve has been reported for some cells heated *in vitro* (Raaphorst *et al.* 1979), but the majority of cell types show a biphasic curve with a second exponential component.

Although cell survival parameters have not yet been obtained for other tissues, the gross tissue response is similarly related to the severity of the heat treatment. Once the threshold time at a given temperature is reached, treatment time–response curves are steep. For example in skin, either rodent (Morris *et al.* 1977; Law *et al.* 1978), porcine or human (Moritz and Henriques 1947), an increase of only 20 per cent in heating time or less than 0.5 °C in temperature increases the probability of thermal necrosis from zero to 100 per cent. As a consequence small inhomogeneities in tissue temperature may lead to marked differences in biological effect. When heating is achieved by immersion in a hot bath correlations between temperature gradients and cell death have been observed in both normal tissue (Hume *et al.* 1979b) and experimental mouse tumors (Hill and Denekamp 1982). It is clear that accurate monitoring of normal tissue temperature during hyperthermia is essential if damage is to be avoided. On the other hand, if tumors are only slightly more sensitive than normal tissues or reach only a slightly higher temperature a therapeutic advantage might be obtained.

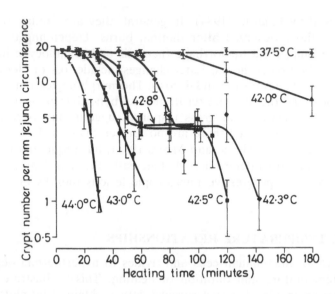

Figure 4: Crypt survival in mouse jejunum after hyperthermia for various times at the temperatures shown. An exteriorised loop of intestine was heated in Krebs–Ringer solution and crypt survival at 1 day after heating was assayed over a 1mm length of jejunum immediately opposite to the mesenteric blood vessel attachments. (Reproduced from Hume and Marigold 1980.)

Comparison of the thermal susceptibilities of different tissues is complicated by the range of heat treatments used and endpoints studied. One approach is to consider the temperatures required to cause injury if heating is maintained for a given time. Temperatures for 60 minutes are given in Table 1. They range from 41.3 °C for mouse testis to 46.8 °C for porcine skin and depend on the endpoint as well as on the tissue.

Some authors have suggested that thermal sensitivity of a tissue depends on the increase from the normal temperature of that tissue rather than on the absolute temperature reached during hyperthermia. One study, in which tumor-bearing mice were acclimatized to different ambient temperatures before heating, supports this suggestion (Hill *et al.* 1983). Table 1, however, does not support a correlation between temperature difference and response. Assuming a normal temperature of about 37 °C for gut and of about ambient, say 22 °C, for ear, the temperature increases required to cause a visible effect in these tissues are very different, being 5 ° and 20 °C respectively. There may, however, be no simple way of relating thermal sensitivities of various tissues. Even if cellular sensitivity were directly related to temperature changes, differences in tissue vascularity may lead to variations in microenvironmental factors, such as oxygenation, nutrient supply and pH, which influence thermal response.

Although thermal susceptibility varies, the relationship between temperature and time required to cause a given level of damage is the same for a

Table 1.
Time–temperature relationships for thermal injury in normal tissues

Tissue	Endpoint	Temperature range (°C)	Transition temperature °C	Time factor for 1 °C Below 42.5 °C	Time factor for 1 °C Above 42.5 °C	Temperature for 60 minutes °C	Authors
Testis of mouse	50 per cent weight loss	39.5–43.8	No transition	–	2.2	41.3	Hand et al. (1979)
Jejunum of mouse	50 per cent crypt loss	42.0–44.5	42.3	8	2.2	42.4	Hume et al. (1979a)
Jejunum of mouse	LD_{50}	43.0–46.0			2.0	42.4	Henle (1982)
Jejunum of hamster	LD_{50}	42.5–44.5			2.0	43.3	Milligan et al. (1984)
Tail of baby rat	Stunting in 5 per cent	42.0–46.0			2.0	43.3	Morris et al. (1977)
Pinna of mouse	Necrosis in 50 per cent	41.5–45.5	42.1	6	2.0	43.3	Law et al. (1978)
Skin of rat	Delay of hair growth	42.0–46.0			1.8	43.4	Okumura and Reinhold (1978)
Tail of baby rat	Necrosis in 50 per cent	41.8–46.0	42.8	6	1.8	43.4	Field and Morris (1983)
Foot of mouse	Loss of toe in 50 per cent	42.5–45.5			2.2	43.6	Overgaard and Suit (1979)
Foot of mouse	Loss of toe in 50 per cent	41.5–46.5	42.5	5	1.9	44.8	Urano et al. (1984)
Foot of mouse	Skin response in 50 per cent	43.5–45.0			2.0	44.8	Robinson et al. (1978)
Foot of mouse	Loss of feet in 50 per cent	43.0–49.0			2.0	45.7	Crile (1963)
Skin of pig and man	Threshold for necrosis	44.0–55.0			2.2	46.5	Moritz and Henriques (1947)

number of normal tissues. A typical isoeffect curve is shown in Fig. 5. Such curves may be described by the relationship:

$$t_2 = t_1 A^{T_1 - T_2}$$

where t = heating time and T = temperature (based on Dewey *et al.* 1977). Values for factor A depend on the temperature range being considered. With the exception of the mouse testis, there is a change in the value of A at a temperature of about 42.5 °C. For temperatures above 42.5 °C if the temperature is increased by 1 °C the heating time has to be reduced by a factor of 2 to give an isoeffect; in other words, 1 °C is equivalent to a factor of 2 in heating time. Below 42.5 °C, a change of 1 °C is equivalent to a factor of about 6 in heating time (Table 1).

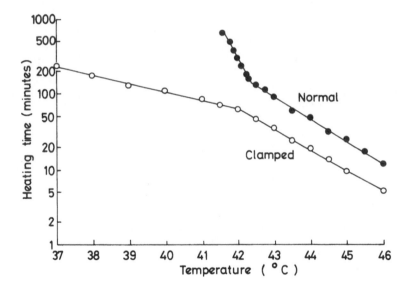

Figure 5: The relationship between heating time and temperature to cause the loss of ten vertebrae in the baby rat tail. Tails were heated by immersion in a water bath under normal conditions (●) or with the tail clamped for at least 20 minutes before and during heating (○). (Redrawn from Morris and Field 1985.)

The activation energy (ΔH) for heat damage can be calculated using the Arrhenius equation:-

$$K \propto e^{-\Delta H/RT}$$

where K = reaction rate of some critical process, taken to be proportional to the reciprocal of the heating time, R = gas constant (8.314 J deg^{-1} mol^{-1}) and T = absolute temperature (Henriques 1947; Westra and Dewey 1971). The values of about 580 kJ/mol in the range of 42.5 °C to 46 °C have been equated with those for protein denaturation, leading to the hypothesis that protein denaturation underlies cell killing by heat. Recent studies, however, have implicated changes in membrane protein complexes rather than protein denaturation *per se* (Leeper 1984). The increased activation energy of about 1500 kJ/mol at temperatures below 42.5 °C may suggest a different mechanism of cell killing. Alternatively, the higher activation energy may be related to the development of thermotolerance during prolonged hyperthermia at low temperatures.

The time–temperature relationships obtained for normal tissues are similar to those for cells *in vitro* (Dewey *et al.* 1977) and experimental tumors *in vivo* (Overgaard and Suit 1979; Nielsen and Overgaard 1982a). This has led to the suggestion that a general time–temperature relationship could be used to define a 'thermal-dose' for treatment planning and comparison of patient response (Field and Morris 1984; Sapareto and Dewey 1984). The approach is to convert thermal exposures at different temperatures to 'equivalent-minutes' at an arbitrarily chosen reference temperature (e.g., 43 °C) using the relationship $t_2 = t_1 A^{T_1 - T_2}$. It is discussed in detail by Sapareto (Chapter 7), but it is pertinent to mention some of the difficulties involved.

(1) The 'inflection temperature', the temperature at which the relationship between time and temperature changes, has been determined only in a limited number of tissues and the results together with those for cells *in vitro* suggest that this temperature may vary.

(2) Although the factor of 2 for temperatures above 42.5 °C is well established, there is considerable uncertainty about the value in the lower temperature range.

(3) In clinical practice there are considerable fluctuations in temperature during any one treatment which might lead to changes in thermal sensitivity resulting from the development of thermotolerance or 'stepdown' sensitization as discussed later. A recent study using the baby rat tail, however, has suggested that variations in thermal sensitivity during such treatment may not be important when the overall tissue response is considered (Field and Morris 1984).

(4) Fractionated hyperthermia is being used to treat patients. It is known that prior heat treatment may reduce subsequent thermal sensitivity and although the slopes of the time–temperature relationship may not change there is evidence that the inflection occurs at a higher temperature in thermotolerant cells and tissues (Bauer and Henle 1979; Law 1979).

(5) Hyperthermia is more likely to be used in combination with more conventional therapy to treat cancer. There is evidence that time–temperature relationships for the interaction between heat and ionizing radiation are similar to those for heat alone although the inflection point is at the lower temperature of 41.5 °C (Hume and Marigold 1985). There are no comparable data for the interaction between heat and other modalities such as chemotherapy and surgery.

D. ENVIRONMENTAL FACTORS AND BLOOD FLOW

Blood flow plays an important role in the clinical application of hyperthermia for two reasons. First, the rise in temperature achieved during local heating depends on heat loss, in which blood flow plays a major role, as well as on the energy supplied by the external heat source. Consequently, regions with low blood flow are likely to reach relatively high temperatures. Second, blood flow determines the microenvironment of the dependent cells and conditions associated with a poor vascular supply, such as hypoxia, nutritional deprivation and low pH, increase thermal sensitivity (Hahn 1974; Overgaard 1976b; Overgaard and Bichel 1977).

The question of whether an inadequate vascular supply is a general feature of all tumors is therefore of crucial importance. Measurement of tumor blood flow in experimental animals and in man has given average perfusion rates which range from less than those of adjacent normal tissues to considerably more (Jain and Ward-Hartley 1984; Song 1984). Such studies, however, do not give information about the distribution of blood flow within a tumor. A limited number of studies in animals have shown that perfusion may be nearly zero in some regions of tumors and several times the average rate in others (Goldacre and Sylven 1962; Shibata and MacLean 1966; Straw et al. 1974; Endrich et al. 1979; Gullino et al 1982). Similar variations have been demonstrated in man (Shibata and MacLean 1966; Beaney 1984).

The importance of regional blood flow in temperature distributions has been demonstrated during local hyperthermia in vivo. In the mouse jejunum the mucosa opposite the mesenteric blood vessels may reach a temperature 1 °C higher than that near the blood vessels (Hume et al. 1979b) and in one rodent tumor, the Walker-256 carcinoma, tumor temperature increased as the perfusion rate decreased (Jain et al. 1979).

The significance of the microenvironment has also been demonstrated; the thermal response of both normal tissues and tumors is increased by occluding the blood supply during heating (Crile 1961; Morris et al. 1977; Baker et al. 1980; Murata et al. 1984; Morris and Field 1985). This is illustrated for the tails of baby rats in Fig. 5. When the blood supply was clamped before and during heating, the increase in thermal susceptibility was equivalent to a factor of about three in heating time for temperatures above 42 °C (Morris et al. 1977; Morris and Field 1985). Although clamping did not change the relationship between time and temperature in the range of 42 °C to 46 °C, it altered the relationship for lower hyperthermal temperatures. As a consequence the effect of clamping increased with decreasing temperature, the increase at 41.8 °C being equivalent to a factor of 6.5 in heating time (Morris and Field 1985). Thus, if poorly vascularized regions of tumor are more susceptible to thermal damage than well perfused areas they will be relatively even more sensitive to heating for long times at low temperatures.

By applying the pressure cuff at various times before the start of heating it has been shown that the very short clamping times required to produce radiobiological hypoxia are insufficient to increase thermal sensitivity. The longer times of clamping before heating which are required to increase thermal sensitivity suggest that changes in pH or nutrient deficiency are the more likely cause of the enhanced heat response than hypoxia per se (Hill and Denekamp 1979; Morris and Field 1979). Tissue pH has been measured by a number of techniques, the most popular being the use of pH electrodes

(Dickson and Calderwood 1983). Measurement with such electrodes, mainly in rodents but including man, has given values for tumor pH ranging from 5.55 to 7.69 compared with the higher values of 7.00 to 8.03 for subcutis and 7.10 to 8.06 for muscle (Wike-Hooley *et al.* 1984; Thistlethwaite *et al.* 1985). As with blood flow, there is considerable intratumor variation in pH. It is generally accepted that electrodes measure the pH of interstitial fluid with an unknown component from damaged cells and blood released from ruptured capillaries. Recent developments in the use of radiotracers with positron emission tomography or of ^{31}P-NMR spectroscopy offer non-invasive methods for estimating apparent pH which includes a component of both intra- and extra-cellular pH. A limited number of studies using such techniques in man have failed to demonstrate a significant pH difference between tumor and normal tissues (Ng *et al.* 1982; Brooks *et al.* 1984; Rottenberg *et al.* 1985). Clearly more data are required before general conclusions about the pH of human tumors can be made. In addition, whether it is intra- or extra-cellular pH which is the important determinant of thermal sensitivity is not established.

Even if tumors are not poorly perfused, differences in the thermal responses of blood vessels in normal tissues compared with tumors could lead to a therapeutic advantage (Bicher *et al.* 1980; Song *et al.* 1980; Dudar and Jain 1984). In general, an increase in blood flow occurs soon after hyperthermia is begun. As the treatment is continued, blood flow increases to a maximum but then decreases, prolonged hyperthermia leading to vascular stasis (Reinhold *et al.* 1978; Bicher *et al.* 1980; Dickson and Calderwood 1980; Song *et al.* 1980; Peck and Gibbs 1983; Reinhold and Berg-Blok 1983a; Stewart and Begg 1983; Dudar and Jain 1984; Karino *et al.* 1984; Lokshina *et al.* 1985; Milligan and Panjehpour 1985). As the heating temperature is increased to 46 °C, the maximum blood flow achieved during hyperthermia increases, the heating time required before reaching the maximum decreases and the time preceding vascular stasis is shortened. Increases in blood flow in animal tumors during moderate hyperthermia are less than those in skin and muscle, presumably leading to a temperature differential (Song *et al.* 1980; Dudar and Jain 1984). In addition, temperatures in the range of 42 °C to 44 °C for 30 to 40 minutes are required for vascular stasis in animal tumors compared with temperatures of more than 46 °C for normal tissues (Bicher *et al.* 1980; Song 1984). Thus temperatures achievable in the clinic are likely to cause vascular stasis and death of dependent cells in tumors without causing marked injury in normal tissues.

After moderate heat treatments, the increased blood flow in normal tissues returns to normal by about 2 days depending on the severity of the initial damage (Song *et al.* 1980; Stewart and Begg 1983; Dudar and Jain 1984). Blood flow in tumors, however, often decreases to values lower than controls after such treatment, although there may be some recovery (Dickson and Calderwood 1980; Song *et al.* 1980; Rappaport and Song 1983; Stewart and Begg 1983). This difference suggests that fractionated hyperthermia given with intervals which allow recovery of normal tissue vasculature but not of tumors could give a therapeutic gain. A study of murine skin and muscle shows that the vascular response of these normal tissues is not impaired during fractionated hyperthermia (Lokshina et al. 1985). The increases in blood flow observed immediately after the last treatment in a fractionated regime were the same as those after a single treatment or even greater. As yet there is no information about the response of tumor vasculature to fractionated treatment.

Changes in microenvironment after hyperthermia have been demonstrated using microelectrodes. Increases and decreases in pO_2 occur in parallel with those in blood flow in both normal tissues and tumors (Bicher et al. 1980; Vaupel et al. 1983). In most experimental studies of pH, that of tumors decreases during hyperthermia whereas that of muscle does not change or may even increase slightly (Bicher et al. 1980; Song et al. 1980; Vaupel et al. 1983). Two published studies in man, however, failed to demonstrate decreases in pH after hyperthermia (Wike-Hooley et al. 1984; Thistlethwaite et al. 1985). This could be due to the lower temperatures achieved in patients rather than an intrinsic difference between experimental animals and man.

These studies, when considered together, support the idea that hyperthermia may give a therapeutic gain. Some areas of tumors may be at low pH and thus more sensitive to heat. The threshold 'thermal dose' for vascular stasis and resultant death of dependent cells is lower for animal tumors than for normal tissues. During local hyperthermia there is a greater increase in blood flow to normal tissues than to tumors so that, even if there is little difference in resting blood flows, the higher temperatures may be achieved in tumors. Hyperthermia may reduce the pH of tumors even further, increasing their susceptibility to further heat treatment.

Care must be taken, however, in extrapolating results of animal studies to the clinical situation. For example, Peck and Gibbs (1983) have shown that the vasculature of jejunum is as sensitive as that of one line of tumors. Furthermore, Dewhirst et al. (1984) have shown that vascular stasis may depend on the rate at which temperature is increased. When temperature was increased at the rate of 0.1 °C/min stasis occurred in tumor vessels at 42.5 °C and in normal tissues at 43.4 ° to 45.2 °C. At the rate of 1 °C/min, however, flow ceased in all vessels at 42.5 °C regardless of their location. There is, however, evidence for vascular collapse in human tumors during hyperthermia (Storm et al. 1979; Bicher et al. 1980; Hofman et al. 1984; Karino et al. 1984) but the thermal dose required to cause stasis is not known.

E.   THERMOTOLERANCE

In clinical practice hyperthermia is being given in multiple fractions, analagous to the use of conventional treatment such as radiotherapy. Knowledge of the biological response to repeated heat treatments is essential for the planning of optimal fractionation schedules. A major consideration is the development of thermotolerance.

Two types of thermotolerance were initially described in cells in vitro. One type is induced by prolonged heating at temperatures below 42.5 °C, the cells becoming more resistant to heat after 3 to 4 hours of hyperthermia (Palzer and Heidelberger 1973; Gerweck 1977; Harisiadis *et al.* 1977; Bauer and Henle 1979). The second type develops after short exposures to hyperthermic temperatures (41 ° to 46 °C) if the cells are returned to normal physiological temperatures. The cells then become increasingly resistant to subsequent heat treatment, the maximum effect being observed after a few hours (Gerner and Schneider 1975; Gerner *et al.* 1976; Henle and Leeper 1976; Henle and Dethlefsen 1978; Nielsen and Overgaard 1979, 1982b; Spiro *et al.* 1982).

Both types of thermotolerance have been observed in tissues *in vivo*. As described earlier, the first has been demonstrated in mouse jejunum as a change in the slope of crypt survival curves following temperatures around 42.5 °C (Hume and Marigold 1980). Its relevance in defining 'thermal dose' was discussed above. The second has been reported for a number of normal tissues including rodent skin (Law *et al.* 1979; Rice *et al.* 1982; Wondergem and Haveman 1984; Urano and Kahn 1985), rodent intestine (Hume and Marigold 1980; Henle 1982; Hume and Marigold 1984; Milligan *et al.* 1984), baby rat tail (Field and Morris 1985), mouse testis (Marigold *et al.* 1985), porcine adipose tissue and skeletal muscle (Martinez *et al.* 1983) but not in canine liver (Prionas *et al.* 1985). It is this type of thermotolerance which is relevant in a fractionated treatment regime.

Figure 6 illustrates thermotolerance in several normal tissues of rodents. In general, thermotolerance increases to a maximum during the first few hours after the priming heat treatment and then decays, normal thermal sensitivity being regained within 1 to 2 weeks. For a given tissue, the degree of thermotolerance is greater and is expressed later following the more severe priming treatment.

Comparison of thermotolerance data is complicated by the number of definitions which have been used. For cells *in vitro* the thermotolerance ratio (TTR) has been defined as the ratio of the D_0 for tolerant cells to that for normal cells (Henle and Dethlesen 1978). This approach has been used for two normal tissues *in vivo*, the mouse intestine, for which crypt survival curves can be constructed (Hume and Marigold 1980), and the mouse testis in which weight loss can be plotted as a function of heating time (Marigold *et al.* 1985). Figure 7 compares the response curves obtained with and without prior heating. The curves are similar to those obtained for cell survival *in vitro* in that prior heat treatment causes a transient decrease in slope (increase in 'thermal D_0'). There is, however, no increase in shoulder as has been reported for some cell lines *in vitro* (Henle and Dethlefsen 1978). At the time of maximum thermotolerance the TTR for jejunum was 14, higher than the values of 4 to 10 reported *in vitro*, but was considerably less, about 2, for the testis. A similar approach has been used for tumors *in vivo* in which the TTR has been defined as the ratio of the slopes of curves relating growth delay to heating time, giving values up to about 8 (Kamura *et al.* 1982; Wheldon *et al.* 1982; Overgaard and Nielsen 1983; Mooibroek *et al.* 1984).

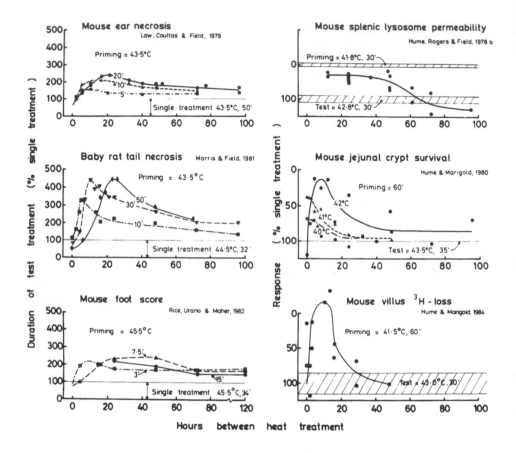

Figure 6: Thermotolerance induced in six normal tissues of mouse heated *in vivo*. In all cases the effects of various priming treatments (D_p) were assessed by giving test treatments at different times after priming. Left-hand diagrams: The duration of the test treatment was varied and the duration (D_t) required to cause a given response in 50 per cent of the animals was expressed as a percentage of the single treatment, D_s. Values greater than 100 per cent indicated thermotolerance. Priming treatments, expressed as a percentage of the single treatment required at the same temperature (D_p/D_s), were 10 per cent (■), 20 per cent (▲), 30 per cent (▼), 40 per cent (●) and 50 per cent (♦). Right-hand diagrams: A chosen test treatment was given and the response was expressed as a percentage of that observed when the test was given without prior hyperthermia. Values less than 100 per cent indicate thermotolerance. (Redrawn from published data as indicated.)

A difficulty in using the TTR to compare thermotolerance in different situations is that it excludes any increases in the shoulder of the cell survival curve and thus may give a misleading impression about the severity of the heat treatment which can be tolerated.

Slope ratios cannot be obtained directly for effects on many normal tissues. Some authors, therefore, express thermotolerance as the ratio of the test treatment required after a priming treatment to cause a given effect (D_t) to that required without priming (D_s) (Law *et al.* 1979; Wondergem and Haveman

1984; Field and Morris 1985).

$$\text{Thermotolerance} = [D_t]/[D_s] \tag{1}$$

An alternative expression includes an additional term for the priming treatment (D_p) (Maher *et al.* 1981; Rice *et al.* 1982; Urano *et al.* 1982).

$$\text{Thermotolerance} = [D_t]/[D_s - D_p] \tag{2}$$

Both of these definitions for thermotolerance *in vivo* have disadvantages. The fact that thermal cell survival curves may have shoulder regions suggests that there may be repair of 'sublethal' heat damage. In definition 1, it is assumed that there is complete repair of thermal injury after the first treatment so that a ratio greater than unity can only be obtained if thermotolerance is induced. This may underestimate thermotolerance, especially if it begins before repair is complete, as is suggested in Fig. 7 for the mouse testis in

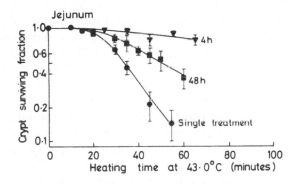

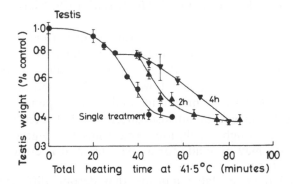

Figure 7: Thermal response curves showing the development of thermotolerance in mouse jejunum and testis. Upper panel: A priming treatment of 42.0 °C for 1 hour, which had no effect on crypt survival, was followed at 4 hours (▼) or 48 hours (■) by test treatments at 43.0 °C. Survival of jejunal crypts is plotted as a function of heating time at 43.0 °C. (Redrawn from Hume and Marigold 1980.) Lower panel: A priming treatment of 41.5 °C for 30 minutes was followed at 2 hours (▲) or 4 hours (▼) by test treatments at 41.5 °C. Testis weight is plotted as a function of total heating time at 41.5 °C. (Redrawn from Marigold *et al.* 1985.) In both panels, a decrease in slope indicates the development of thermotolerance.

which there is a decrease in slope of the response curve without full recovery of the 'shoulder' region (Marigold *et al.* 1985). Definition 2, however, would give a value greater than unity if there were only repair, thus implying non-existent thermotolerance.

There are also empirical differences between the two expressions. The first, which gives a direct measure of the severity of the heat treatment required to cause an effect in thermotolerant tissue, increases to a plateau as the priming treatment is increased (Law *et al.* 1979; Field and Morris 1985). The second, which approximates to the ratio of slopes of survival curves used *in vitro*, continues to increase suggesting that the few cells which survive severe heat treatment are extremely thermotolerant (Maher *et al.* 1981; Rice *et al.* 1982; Urano *et al.* 1982; Field and Morris 1985).

The *in vivo* development of thermotolerance after an acute treatment depends on the temperature as well as on the duration of heating (Hume and Marigold 1980; Law 1981; Field and Morris 1985). Thermotolerance induced by different temperatures has been compared by expressing the duration of a priming treatment at a given temperature (D_p) as a fraction of the heating time (D_s) required as a single treatment at the same temperature to cause direct thermal injury (Nielsen and Overgaard 1982b; Field and Morris 1985). In the baby rat tail both the maximal thermotolerance and the time interval required before the maximum was reached depended on the fractional priming treatment (D_p/D_s) rather than the temperature in the range 42 °C to 45 °C (Field and Morris 1985). Clamping the tail, thus increasing its sensitivity, had no effect on the relationship between D_p/D_s and thermotolerance (Field and Morris 1985). These results support the suggestion made for cells *in vitro* that it is the level to which cell survival is reduced, rather than the heat treatment used to achieve it, which determines the degree of thermotolerance induced (Henle *et al.* 1979; Nielsen and Overgaard 1982b).

In view of the different thermal sensitivities of normal tissues (Table 1) it seems reasonable to use the concept of fractional priming treatment to compare thermotolerance in different tissues. Results for similar fractional priming treatments (D_p/D_s) are shown in the left hand diagrams of Fig. 6. For a given D_p/D_s both the magnitude and timing depends on the tissue and endpoint studied, suggesting intrinsic differences between the various tissues studied. Thus, although conclusions concerning the degree and duration of thermotolerance may be made for a given tissue it is difficult to generalize from tissue to tissue.

The limited data available indicate that despite different capacities to develop thermotolerance there is a general relationship between the heating time and temperature required for its induction. This relationship is similar to that for direct thermal injury; above about 42.5 °C a change of 1 °C is equivalent to a factor of 1.6 to 2 in time and below 42.5 °C to a factor of about twice the lower value (Law 1981; Nielsen and Overgaard 1982 a,b; Morris and Field 1985). The relationship between time and temperature for thermal injury in tissues in which thermotolerance has been induced by prior heating is also similar to that for previously unheated tissues; the complete curve including the inflection point, however, is shifted to higher temperatures (Bauer and Henle 1979; Law 1979). This similarity suggests that the same mechanisms underly thermal damage in both previously unheated and thermotolerant cells.

The extent to which thermotolerance in tissues is due to a decrease in intrinsic thermal sensitivity or is secondary to vascular changes which modify the cellular environment is not clear. The similarities between thermotolerance in tissues *in vivo* and cells *in vitro* indicate that it is a cellular phenomenon. The mechanisms which underly thermotolerance in cells, however, are unknown. Recent studies have implicated an important role of the so-called heat shock proteins which are synthesised after a thermal insult. Current ideas concerning cellular mechanisms of thermotolerance are reviewed elsewhere (Burdon 1984; Li and Hahn 1984).

In vivo, however, changes in vascular perfusion may alter the cellular response. In some situations heat-induced increases in blood flow may augment heat loss and nutrient supply, thus making later treatments less effective. In others, vascular damage could reduce heat loss so that subsequent local heating would cause a greater increase in temperature and lead to nutrient deficiency with a consequent reduction in pH. These factors would make cells more sensitive to subsequent heating and the reduced pH would inhibit the further development of thermotolerance (Nielsen and Overgaard 1979; Goldin and Leeper 1981 a,b).

Indirect evidence that tissue environment may modify thermotolerance *in vivo* has been obtained for the baby rat tail (Morris and Field 1985). As illustrated earlier (Fig. 5) occluding the blood supply eliminated the transition in the curve relating heating time and temperature, a result consistent with *in vitro* studies which show that low pH inhibits the development of thermotolerance during heating at temperatures below 42.5 °C. It may not be possible, however, to investigate the role of pH in the development of thermotolerance after acute heat treatment *in vivo* using existing techniques. Studies *in vitro* have shown that a low pH must be maintained during the interval between fractions for any reduction in thermotolerance to be observed (Nielsen and Overgaard 1979; Gerweck *et al.* 1983a). It is clearly impractical to maintain a tissue clamp for the time required for the development of thermotolerance *in vivo* (Fig. 6). Other methods of lowering tissue pH, such as glucose overload or drug treatment, have been considered but the results are inconclusive (Dickson and Calderwood 1983).

If regions of low pH do occur in tumors it is likely that such tumors will develop less thermotolerance than normal tissues. A few studies, however, have shown that thermotolerance in animal tumors lies within the range of results for normal tissues (Maher *et al.* 1981; Kamura *et al.* 1982; Urano *et al.* 1982; Wheldon *et al.* 1982; Overgaard and Nielsen 1983; Mooibroek *et al.* 1984). It has also been suggested that, since cells which are hypoxic, and presumably in tissues at low pH, are resistant to irradiation, the tumor cells which survive radiotherapy may be at a low pH. Nielsen *et al.* (1983), however, estimated thermotolerance in such a population after irradiation and found no reduction in tolerance compared with that of the unirradiated tumors. Thus the experimental studies to date have given no clear indication that tumors contain cells which have a reduced capacity to develop thermotolerance and there is no basis for optimism that differences in the development of thermotolerance in normal tissues and tumors may lead to a therapeutic gain.

The decay of thermotolerance has received less attention than its development, somewhat surprisingly in view of its importance in the response to multifractionated heat treatment. In the baby rat tail, the fractional rate of decay of tolerance is independent of the priming treatment, the maximal tolerance induced by any one treatment falling to 10 per cent in about 5 days (Field and Morris 1985). There is inadequate data to say whether or not this relationship holds for other tissues. The evidence available, however, indicates that decay varies considerably from tissue to tissue (Law et al. 1979; Hume and Marigold 1980; Rice et al. 1982; Milligan et al. 1984; Field and Morris 1985). This variation may be related to differences in tissue turnover times. It is generally accepted that heat-induced thermotolerance is not heritable but is lost at mitosis so that as cells divide the number of thermotolerant cells decreases (Henle and Dethlefsen 1978). Studies of cells in vitro show that although the rate of induction of thermotolerance does not depend on proliferative status, its decay is more rapid in proliferating than in plateau phase cells (Hahn 1982; Gerweck et al. 1983; Gerweck and Delaney 1984). These latter results suggest that slowly dividing normal tissues may retain thermotolerance longer than more rapidly proliferating tumors, thus giving a therapeutic advantage in fractionated hyperthermia. Data for the mouse jejunum, however, show the development and decay of thermotolerance in the proliferative epithelium of the crypts to be similar to those in the non-proliferative villus compartment (Hume and Marigold 1984). Nor does the limited information available suggest a marked difference in the decay of thermotolerance in mouse tumor and skin. Detailed studies of thermotolerance in slowly dividing tissues (e.g., muscle, lung, kidney, brain) have not been reported.

ated course of hyperthermia has been considered in a limited number of studies. Henle et al. (1979) have reported that repeating treatments at 45 °C and returning cells to 37 °C between treatments induced a relatively constant level of thermotolerance in CHO cells in vitro; they suggested that thermotolerance reaches an absolute maximum. Similarly, Nielsen and Overgaard (1985) using L1A2 tumor cells in vitro, showed that, although thermotolerance could be built up during a fractionated regime if the interfraction intervals were not long enough to allow the complete development of tolerance, there was a maximum thermotolerance which would not be exceeded.

Equal heat treatments have been used in some in vivo investigations, analogous to studies of effects of fractionated radiation. Urano et al. (1980) found that fractions of equal size did not cause the same effect throughout a course of treatment and suggested that more thermotolerance developed after the later fractions than after the early ones. Overgaard *et al.* (1980) also reported an increase in recovery ratios with treatment number suggesting that thermotolerance was built up during the course of fractionated treatment.

When only equal heat fractions are used it is not possible to distinguish repair of thermal damage (a recovery of the shoulder) from the development of thermotolerance (an increase in D_0 with or without an increase in shoulder). More recently, some authors have given a variable test heat treatment to measure thermotolerance after different multifraction heat regimes *in vivo* (Overgaard and Nielsen 1983, 1984; Law *et al.* 1984a). These studies indicate that there is a maximal degree of thermotolerance which cannot be exceeded. The development of thermotolerance, however, depends on the interval between fractions. This is illustrated for the mouse ear in Fig. 8 which shows the effect of increasing the interval between two fractions. If the second fraction was given 4 hours after the first when thermotolerance was developing, complete development was delayed for 4 hours but the maximal value and decay of tolerance were the same as after a single treatment. If the second fraction was given at 24 hours when thermotolerance was maximal, its value was maintained for 24 hours before decaying. If the second fraction was given at 72 or 168 hours when thermotolerance was decaying, there was a further increase to the maximum followed by decay. In addition, five fractions given at 24 or 72-hour intervals induced thermotolerance of the same magnitude and time course as that induced by two fractions separated by the appropriate interval (Law 1985).

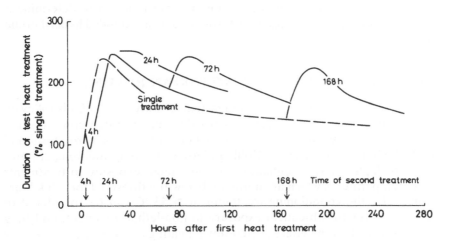

Figure 8: Thermotolerance induced in the mouse ear by fractionated hyperthermia. Thermotolerance was assessed by giving a variable test treatment at 43.5 °C and expressing the duration of the test required to cause necrosis in 50 per cent of the ears as a percentage of that required without prior heating. Values greater than 100 per cent indicate thermotolerance.
Dotted line: Thermotolerance after a single treatment of 20 minutes at 43.5 °C.
Solid lines: Thermotolerance after two treatments of 20 minutes at 43.5 °C separated by 4, 24, 72, or 168 hours. (From M.P. Law, unpublished.)

It has been suggested that thermotolerance induced by a single priming treatment may be increased by giving additional heat treatments which would not be tolerated by normal cells. In the mouse ear, however, after thermotolerance had been induced by a single treatment, giving up to nine additional treatments, each of which would cause necrosis in ears which had not received prior hyperthermia, did not increase thermotolerance. In fact, thermotolerance decreased as the number of severe treatments was increased from 4 to 9 (Law *et al.* 1984a).

The results of *in vivo* studies of fractionated hyperthermia could be explained if each treatment killed a proportion of cells but induced an increasing degree of thermotolerance (Urano *et al.* 1982; Law *et al.* 1984b). Alternatively, after the first treatment which kills some cells but induces tolerance, subsequent heating could maintain a constant degree of thermotolerance without killing a significant proportion of cells (Law *et al.* 1984b, 1985b). Law *et al.* (1984b, 1985b) have estimated the survival of epithelial cells in the mouse ear after various fractionated heat regimes by giving a test radiation treatment. The results depended on the magnitude of the heat treatments used but they indicate that each fraction (when given at maximal tolerance) killed few, if any, cells but maintained thermotolerance. Results obtained for L1A2 tumor cells *in vitro* are consistent with this conclusion (Nielsen and Overgaard 1985).

An alternative explanation is that heat stimulates repopulation. The development of thermotolerance after a single treatment is probably too rapid to be explained by repopulation. Repopulation, however, may be important in fractionation regimes which extend over several weeks (Urano *et al.* 1980, 1982; Rice *et al.* 1982). Studies of cell kinetics are required to determine if cell death is balanced by increased cell turnover during extended hyperthermia regimes.

F.   THERMOSENSITIZATION BY STEP-DOWN HEATING

Hyperthermal treatments in which a high temperature (> 43 °C) is followed by a low temperature (< 43 °C) are described by the term step-down heating. The effect of step-down heating was first demonstrated *in vitro* (Joshi and Jung 1979; Henle 1980; Jung and Kolling 1980). Cell killing after temperatures of 43 °C or higher may be enhanced by subsequent exposure to treatments at lower temperatures which are normally below the threshold for cell killing. Alternatively, the conditioning treatment at > 43 °C can be considered to sensitize the cells to subsequent exposure to non-lethal treatment, and Jung (1982) has described the effect of step-down heating as an increase in cellular sensitivity (a decrease in D_0) to treatment at temperatures below 43 °C when the treatment is preceded by hyperthermia at 43 °C or higher. *In vitro*, the degree of sensitization depends on the severity of the pretreatment and the effect is lost if the cells are maintained at 37 °C for 2 hours before heating at the lower temperature (Jung 1983).

Thermosensitization of normal tissues by step-down heating has been demonstrated in a limited number of studies (Miyakoshi *et al*. 1983; Reinhold and van den Berg-Block 1983b; Urano and Kahn 1983; Field and Morris, pers. comm.; Hume and Marigold, pers. comm.) although absence of the effect has been reported in mouse foot (Henle and Dethlefsen 1982). In the two more detailed studies of mouse skin, various treatments at a 'high' temperature were followed immediately by a constant treatment at a 'low' temperature (Miyakoshi *et al*. 1983; Urano and Kahn 1983), as illustrated in Fig. 9. In such experiments the skin response to the step-down sequence was greater than that after the 'high' temperature treatment given alone. This effect has been described by an enhancement factor defined as the ratio of heating times at the 'high' temperature required to cause a given skin response with or without immediate heating at the 'low' temperature (Miyakoshi *et al*. 1983; Urano and Kahn 1983). This factor depends on the severity of the 'low' temperature treatment but it does not distinguish between true sensitization, as demonstrated *in vitro* by a decrease in D_0, and a purely additive effect of the two consecutive treatments.

The second 'low' temperature (T_2) treatment may be converted to approximate equivalent heating time at the 'high' temperature (T_1) and total treatment time at T_1 considered. Experiments both *in vitro* (Henle and Dethlefsen 1980) and *in vivo* (Urano and Kahn 1983; Field and Morris 1984) indicate that the transition in the heating time-temperature isoeffect curve is lost when treatment is immediately preceded by 'high' temperature treatment, 1 °C being equivalent to a factor of 2 in the range 40 °C to 46 °C. Thus, 60 or 120 minutes at 42 °C is equivalent to a maximum of 15 or 30 minutes at 44 °C respectively. In Fig. 9, the observed responses to step-down heating are greater than those expected if the heating time at 44 °C were increased by 15 or 30 minutes, indicating true sensitization. If there were an inflection at 42.5 °C there would be an even greater difference between observed and expected response.

Sensitization to the second treatment has been clearly demonstrated in a study using the mouse intestine (Hume and Marigold, pers. comm.). As shown in Fig. 10, a constant 'subthreshold' treatment at 43 °C was given to an isolated loop of jejunum before variable treatments at 42.5 °C. Prior treatment at 43 °C for 20 minutes reduced the shoulder and increased the slope of the crypt survival curve obtained for heating at 42.5 °C. The decrease in shoulder could be accounted for by a simple addition of treatments, whereas the increase in slope represented true sensitization to heating at 42.5 °C.

As with cells *in vitro*, thermosensitization by step-down heating *in vivo* increases if the duration of the initial 'high' temperature treatment is increased (Miyakoshi *et al*. 1983; Urano and Kahn 1983). The effect, however, is lost as the interval between the two treatments is increased (Miyakoshi *et al*. 1983; Urano and Kahn 1983). In mouse skin the interaction remained maximal for an interval of about 6 hours but was lost by about 15 hours (Urano and Kahn 1983). In contrast, in the intestine step-down sensitization was more rapidly lost and thermotolerance was observed when the interval was 3 hours

M. P. Law

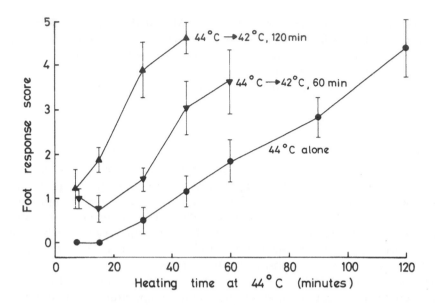

Figure 9: The response of the mouse foot to step-down heating. Immediately after heating at 44 °C for various times feet were further exposed to 42 °C for 60 minutes (▼) or 120 minutes (▲). For the control treatment, feet were heated at 44 °C alone (●). Response was assessed using an arbitrary numerical score and maximum scores are plotted as a function of time at 44 °C. Enhancement factors, calculated for a response of 3, were 2 for the 60-minute treatment at 42 °C and 4 for the 120-minute treatment. (Redrawn from Miyakoshi *et al.* 1983.)

(Fig. 10). The degree to which the differences in the duration of the sensitizing effect is attributable to physiological changes in heated tissues *in vivo* is unknown.

The rapid loss of the effect of step-down heating has led both Henle (1980) and Jung (1982) to interpret the sensitization in terms of an interaction of 'sublethal damage' following a primary treatment with subsequent 'low' temperature treatment such that the lesions become lethal. Studies which show that sensitization can not be induced by step-down heating in thermo-tolerant cells or tissues, however, suggest that the increased sensitivity reflects inhibition of the development of thermotolerance (Hahn 1982; Jung 1982; Nielsen *et al.* 1982; Urano and Kahn 1983). The observation, both *in vitro* (Henle and Dethlefsen 1980) and *in vivo* (Field and Morris 1984), that prior heat treatment at a 'high' temperature removes the transition in the time–temperature isoeffect curve has been considered to be evidence supporting the elimination of thermotolerance. Against this hypothesis are the observations that (1), biphasic survival curves, indicative of the development of ther-motolerance during prolonged heating at 'low' temperatures, can be obtained after prior conditioning at a 'high' temperature (Henle 1980); (2), the development of thermotolerance in skin (Urano and Kahn 1983) and intestine (Hume and Marigold, pers. comm.) is not affected by step-down heating;

and (3) sensitization by step-down heating has been reported for L1A2 tumor cells which show neither a biphasic survival curve nor an inflection in the time-temperature relationship for a single heat treatment (Nielson *et al.* 1982). Studies using L1A2 tumor cells have shown that step-down heating caused a break in the time-temperature relation giving an activation energy of 238 kJ/mole for temperatures below 42.5 °C. (Nielsen *et al.* 1982). This result suggests that the mechanism of cell killing under step-down heating conditions differs from that in single-heated or thermotolerant cells.

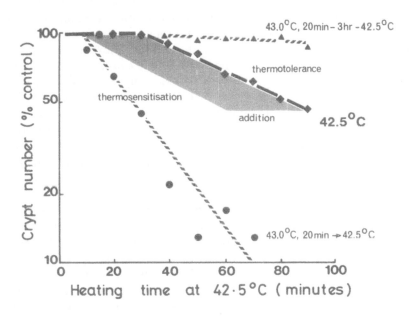

Figure 10: The effect of step-down heating in the jejunum of mouse. Crypt survival was assessed after heating an exteriorized loop of jejunum at 42.5 °C (♦) or at 43 °C for 20 minutes followed immediately (●) or 3 hours later (▲) by 42.5 °C. The shaded area shows the heat treatment of 20 minutes at 43 °C expressed as the equivalent time at 42.5 °C. (From S.P. Hume, unpublished.)

The clinical importance of step-down sensitization during a fluctuating heat treatment was discussed above. Few studies have considered the effect of step-down heating in tumors. One indicates enhanced damage in a transplanted tumor in mouse by step-down heating, without any apparent additional damage to normal tissues of the foot (Henle and Dethlefsen 1982). Reinhold and van den Berg-Blok (1983b), however, failed to demonstrate sensitization of the microcirculation of an experimental tumor in mouse, although they did record an effect on the surrounding normal tissue or tumor bed. In a more detailed study Urano and Kahn (1983) report that step-down sensitization in a mouse fibrosarcoma was similar in degree and duration to that in skin and comparable results have been obtained by Lindegaard and Overgaard (1985). It seems unlikely, therefore, that a step-down sequence of treatment will give a therapeutic gain.

G.   RE-TREATMENT BY HYPERTHERMIA

The thermal response of tissues which have already received a complete course of treatment, either radiotherapy or chemotherapy, is of interest since hyperthermia is being used to treat tumors which recur after the more conventional clinical treatments. Clinical experience of treating superficial tumors with hyperthermia has indicated that prior radio- or chemotherapy has little effect on the response to subsequent hyperthermia (Marmor and Hahn 1978; Marmor et al. 1979). These results have led Marmor and Hahn (1979) to suggest that heat may be particularly useful in treating recurrences for which the prognosis for conventional therapy is poor. There is evidence, however, that an earlier treatment may lead to an unexpected response to heat. For example, whole-body hyperthermia has caused acute myelopathy in a few patients who had received neurotoxic drugs or local irradiation of the spinal cord (Barlogie et al. 1979; Parks et al. 1979; Douglas et al. 1981). The report of Douglas et al. (1981) suggests that the effect of prior treatment is transient. The patients, three out of twelve, who developed myelopathy after hyperthermia had received spinal cord irradiation and chemotherapy less than 2 months earlier. None of the 60 patients given hyperthermia more than 2 months after initial radiotherapy treatment developed acute neurological symptoms.

The interpretation of the clinical results is complicated by the fact that the initial treatment may include radiotherapy, chemotherapy and even surgery. Although there are no published experimental studies of the long-term effects of chemotherapy or surgery on thermal response there is limited information about those of irradiation. In general, thermal response in rodents is increased when heat is given after the expression of acute radiation injury (Hume and Marigold 1982; Law and Ahier 1982; Baker et al. 1983; Wondergem and Haveman 1983; Law et al 1984c, 1985a; Hume and Marigold 1986). In skin of the foot, irradiation given 5 or 13 weeks earlier increased the duration of the thermal response without a marked increase in the maximal skin response (Baker et al. 1983; Wondergem and Haveman 1983). Whereas the erythema or desquamation induced by heat alone had healed by 20 days after heating, with prior radiotherapy the thermal response persisted for up to 120 days. In the skin of the ear, however, prior irradiation had no effect on the time course of the thermal response but it did increase thermal sensitivity, as measured by the heating time at a given temperature required to cause thermal necrosis (Law and Ahier 1982).

Increases in thermal response depend on the interval between irradiation and heating, as illustrated in Fig. 11. In the intestine, thermal susceptibility was maximal at 2 weeks after a single dose of 9 Gy but had returned to normal by 12 weeks. In contrast, thermal susceptibility of ear skin was increased for up to a year after doses of 17–20 Gy. This apparent difference may be related to radiation dose. In the ear there was a threshold radiation dose ($\sim$ 12 Gy) below which no changes in thermal sensitivity could be detected and there was evidence of a return to normal thermal sensitivity after radiation doses just above this threshold (Fig. 11).

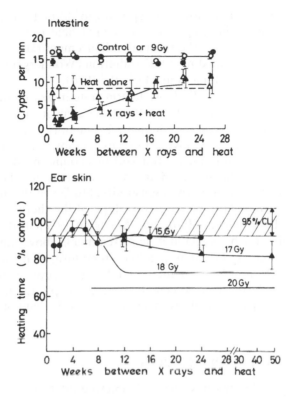

Figure 11: The effect of prior irradiation on the thermal response of jejunum and ear of mouse. Upper panel: An exteriorized loop of jejunum was heated at 43.0 °C for 35 minutes at various times after 9 Gy (▲) or 20 Gy, 4 fractions/4 days (■) of X-rays and crypt loss assayed one day later. Also shown are crypt counts in the jejunum of age-matched mice which received heat alone (△), radiation alone (●) or were untreated (○). (Redrawn from Hume and Marigold 1982.) Lower panel: Ears were heated at 43.5 °C at various times after 15 Gy (●), 17 Gy (▲), 18 Gy or 20 Gy and the heating time required to cause necrosis in 50 per cent of the irradiated ears (NT_{50}) was expressed as a percentage of that required in age-matched controls. A decrease in NT_{50} indicates an increase in thermal susceptibility. (Data from Law and Ahier 1982; Law *et al.* 1984c, 1985a.)

There is little experimental information about the effects of fractionating either initial therapy or re-treatment. Studies of the mouse intestine and ear show that single or fractionated doses of X-rays which cause comparable acute radiation responses have similar effects on subsequent thermal sensitivity (Hume and Marigold 1982; Law *et al.* 1985a). As regards fractionating the heat treatment, prior irradiation has been shown to have no effect on the development of thermotolerance (Wondergem and Haveman 1983).

It has been suggested that the long-term increase in thermal susceptibility may be related to the increase in radiosensitivity seen in several rodent tissues at various times after prior irradiation (Law and Ahier 1982). Residual X-ray damage may interact with subsequent thermal injury to increase apparent thermal sensitivity, but with time this residual radiation damage may be repaired. This interaction, however, appears to differ from that suggested to explain the radio-enhancing effect of a mild heat treatment since prior irradiation does not increase the thermal enhancement ratios obtained when the re-treatment is combined heat and radiation (Wondergem and Haveman 1983; Law *et al.* 1985a; Hume and Marigold 1986).

Whereas the experimental findings demonstrate increased thermal susceptibility after prior irradiation, it is difficult to evaluate the effect of prior treatment on the thermal response in man. The majority of patients included in early trials have already received previous therapy and in many cases the probability of superficial lesions is reduced by cooling the skin. In addition, the hyperthermia treatments used in the clinic may not be sufficient to demonstrate any change in thermal sensitivity. Studies of the mouse intestine show that prior irradiation increases the slope of the thermal crypt survival curve but not the shoulder (Hume and Marigold 1982) so that a change in sensitivity would only be detected if heat treatments above the threshold for thermal injury were given. It is possible, therefore, that more aggressive heat therapy, which will be made possible by improvements in heating technology, may lead to unexpectedly severe complications if used after previous radiotherapy, particularly in deep organs which cannot be adequately cooled.

H.   WHOLE-BODY HYPERTHERMIA

Whole-body hyperthermia has received renewed interest as a treatment for patients with metastatic disease and as a method for heating deep-seated tumors which as yet cannot be treated by localized hyperthermia (Bull 1983; Parks and Smith 1983). A disadvantage of this treatment is that most mammalian species are unable to tolerate whole-body heating at temperatures above 43 °C for any longer than a few minutes; heating for several hours at tolerable temperatures (< 42 °C) is required for tumor-cell destruction. In addition, the clinical effectiveness of such treatment depends solely on any difference between tumor and normal tissue sensitivity, as no temperature differential is obtained. The most promising application of systemic hyperthermia, however, will probably be in combination with appropriate chemotherapeutic agents (Bull 1984).

It is well known that heat stress places extreme demands on the cardiovascular and respiratory systems, sweating functions, salt and water balance, renal function and neurosystem control (Ladell 1964; Lee 1964). The majority of the methods used to induce whole-body hyperthermia in the clinic, however, require general anesthesia and during treatment patients receive ventilation and fluid support which counteract some of the physiological responses to systemic hyperthermia (Cronau *et al.* 1984; Milligan 1984). The systemic temperature which may be achieved during therapy is limited to 42 °C by the tolerance of liver and brain as described by Pettigrew *et al.* (1974) and by the development of disseminated intravascular coagulation (Ludgate *et al.* 1976). Whether the temperature tolerance is also limited by response of other organs such as lung, heart or kidney has not been established.

Experimental studies of the effects of whole-body hyperthermia in small animals have been limited by the difficulties encountered in attempts to maintain core temperature in the therapeutic range. Systemic hyperthermia has been achieved by water bath heating under anesthesia but rodents are particularly sensitive to this technique. For example, in rats, 41.5 °C for 60 minutes reduced survival to 10 per cent (Dickson and Ellis 1974) and 42 °C for 15 minutes was lethal (Bland *et al*. 1982). A factor in these poor survival rates may have been the barbiturate anesthesia, which is associated with cardiac and respiratory depression. More recently Kapp and Lord (1983) have reported that water bath heating of rats using gaseous anesthesia (halothane 1 per cent: oxygen 99 per cent) improves survival; 42.5 °C for 25 minutes only reduced animal survival to 50 per cent.

Mice are as sensitive to whole-body hyperthermia under general anesthesia as are rats. Using 2450 MHz irradiation, a systemic temperature of 43.5 °C for 20 minutes reduced survival to 30 per cent (Tsubouchi *et al*. 1984) although a temperature of 42 °C for 35 minutes was tolerated (Nakayama and Nakamura 1984). Survival may be improved by replacing the fluid lost during exposure to hyperthermia but even then the maximum treatment that could be tolerated, as a lower body exposure, by mice was only 42 °C for 1 hour (Hume *et al*. 1978).

Unanesthetized animals are more resistant to systemic hyperthermia. For example, rats can tolerate 41.2 °C to 41.8 °C induced by a radiant heat apparatus for 90 minutes, without showing any clinical signs of distress, provided an intraperitoneal injection of saline is given to prevent dehydration (Neville *et al*. 1984). Mice have also been heated in warm air to temperatures of 41 °C for 50 minutes without adverse effects (Honess and Bleehen 1985).

The majority of animals are lost during the first day after treatment, many dying within a few hours. Studies of mice indicate that thermal damage in the intestine is a major factor in these acute deaths, although edema of lungs has also been reported (Tsubouchi *et al*. 1984; Nakayama and Nakamura 1984). However, the factors which limit whole-body hyperthermia in man, i.e., liver damage, myelopathy and coagulation defects, have not been studied extensively in animals. Local hyperthermia has indicated that the liver is particularly sensitive, the threshold for damage being 41 °C for 30 minutes in dogs (Prionas *et al*. 1985) and 42 °C for 15 minutes in rats (Adam *et al*. 1985). Whereas changes in serum levels of liver enzymes and bilirubin, indicating liver damage, have been reported after whole-body hyperthermia in man (Pettigrew *et al*. 1974; van der Zee *et al*. 1983), information about liver function after heating animals is limited. Depressions in bile secretion and increases in the release of the enzyme aspartate aminotransferase has been observed after heating isolated rat liver at 42 °C (Wynne *et al*. 1984). Increases in serum glutamate oxaloacetate transaminase (SGOT) and lactate dehydrogenase (LDH), but not bilirubin, have also been reported after local hyperthermia to the liver of rabbits (Boddie *et al*. 1985).

One study designed to investigate neuropathy in rats failed to demonstrate any neurological signs after systemic hyperthermia at 42 °C for 90 minutes, induced by a radiant heat device (Neville *et al.* 1984). Changes in platelet counts, suggesting low grade disseminated intravascular coagulation, however, have been reported in mice (Nakayama and Nakamura 1984). After 42 °C for 35 minutes platelet number reached a minimum at 3.5 hours and then gradually increased to more than twice the normal value by 3 days. It was suggested that the thrombocytosis arose to offset a deficiency of platelets induced by aggregations and adhesions. A transient depletion of hemopoietic stem cells has been reported after local hyperthermia at 42 °C for 1 hour to the tibia of mice and more severe treatments caused a further depletion to less than 50 per cent control values by 3 months (Werts *et al.* 1984). Whether or not whole-body hyperthermia causes a chronic decline in hemopoietic sufficiency requires investigation.

These initial experimental studies suggest that the prolonged treatment required for tumor control is not attainable in small animals, especially under anesthesia, making the relevance of rodent studies to the clinical problems of whole-body hyperthermia questionable. New techniques, such as the radiant heat device (Robbins *et al.* 1985; Neville *et al.* 1985), may offer ways of maintaining systemic hyperthermia with minimal complication, thus stimulating interest in experimental studies.

I. CONCLUSIONS

There are several reasons why hyperthermia may be useful in the treatment of cancer. Solid tumors which have an inadequate vascular supply may reach higher temperatures than adjacent normal tissue during local heating. Some of the cells in such tumors may be deprived of nutrients and at low pH; these conditions increase thermal sensitivity. Increases in tumor blood flow during hyperthermic treatments which are sufficient to cause an hyperemic response are less than those in normal tissues and lower 'thermal doses' are required for vascular collapse in tumors than in normal tissues. These factors should give a therapeutic advantage for hyperthermia.

Thermal damage to tissues is expressed within a few hours of heating, consistent with intermitotic cell death. Stromal damage is especially important in hyperthermal injury, but subcellular events leading to cell death after heating are not established.

Thermal damage in tissues depends on both the temperature and duration of heating. Once a threshold is reached, a small increase in temperature or heating time causes a marked increase in the probability of damage, emphasising the clinical need for careful thermal dosimetry. Although the thermal susceptibility of tissues varies, there is a general relationship between temperature and heating time required for a given response. This relationship has been used to express clinical treatments at different temperatures as an equivalent heating time at 43 °C, but the validity of this approach is not established.

Hyperthermia induces transient resistance (thermotolerance) to subsequent heating. Both the magnitude and time course of thermotolerance depend on the heat treatment used to induce it. Although the ability of various tissues to develop thermotolerance differs, the relationship between temperature and heating time for its induction is similar to that for direct thermal injury. Thermotolerance induced in experimental tumors is similar to that in normal tissues. In view of this, and of the variation between normal tissues, it may be advisable to give clinical treatments at intervals which allow the decay of thermotolerance.

Step-down heating increases the thermal sensitivity of normal tissues to 'low' temperatures (< 43 °C). It has a similar effect in tumors so it is unlikely that a step-down sequence will give a therapeutic gain.

Animal experiments indicate that previously irradiated skin and intestine are more susceptible to thermal damage than untreated controls. Although hyperthermia is being used without complications to treat superficial tumor recurrences, it is possible that more aggressive heat therapy than is used at present may lead to unexpected responses if used after previous treatment.

Whole-body hyperthermia offers a method of treating systemic disease but is limited in man to 42 °C by the tolerance of liver and brain. Rodents are particularly sensitive to whole-body heating under general anesthesia and consequently may not be useful models for the clinical problems of systemic hyperthermia.

Clinical studies indicate that human tumors can be treated without unacceptable complications. Regression of tumors has been reported after heat alone, but the greatest potential for the use of hyperthermia in the treatment of cancer is in combination with radiotherapy or chemotherapeutic agents.

ACKNOWLEDGEMENTS

I would like to thank Dr S. P. Hume, Dr A. Michalowski and Mr R. G. Ahier for their invaluable discussions of the manuscript and Dr S. P. Hume for allowing me to include her unpublished data.

REFERENCES

Adam, R., Poggi, L., Houssin, D., Capron, M., Morin, J., Jijou, M., Miramand, J.C., Szekely, A.M. and Bismuth, H. (1985). Effects of hyperthermia on normal or neoplastic rat liver. *Eur. Surg. Res.* **17**, 310–319.
Badylak, S.F., Babbs, C.F., Skojac, T.M., Voorhees, W.D. and Richardson, R.C. (1985). Hyperthermia-induced vascular injury in normal and neoplastic tissue. *Cancer* **56**, 991–1000.
Baker, G.M., Waas, A.N.C. and Wright, E.A. (1980). The influence of ischaemia on hyperthermic damage to mouse tail. *Int. J. Radiat. Biol.* **37**, 109–114.
Baker, D.G., Sager, H., Constable, W. and Goodchild, N. (1983). The response of previously irradiated skin to combinations of X-radiation and ultrasound-induced hyperthermia. *Radiat. Res.* **96**, 367–373.
Barlogie, B., Corry, P.M., Yip, E., Lippman, L., Johnston, D.A., Khalil, K., Tenczynski, T.F., Reilly, E., Lawson, R., Dosik, G., Rigor, B., Hankenson, R. and Freireich, E.J. (1979). Total body hyperthermia with and without chemotherapy for advanced human neoplasms. *Cancer Res.* **39**, 1481–1489.
Bauer, K.D. and Henle, K.J. (1979). Arrhenius analysis of heat survival curves from normal and thermotolerant CHO cells. *Radiat. Res.* **78**, 251–263.
Beaney, R.P. (1984). Positron emission tomography in the study of human tumors. *Semin. Nucl. Med.* **14**, 324–341.

Bicher, H.I., Hetzel, F.W., Sandhu, T.S., Frinak, S., Vaupel, P., O'Hara, M.D. and O'Brien, T. (1980). Effects of hyperthermia on normal and tumor microenvironment. *Radiology* **137**, 523–530.

Bland, K.I., Hodge, K.M., Bean, R.J., Akin, J.R. and Fry, D.E. (1982). Effects of systemic hyperthermia on the growth patterns of experimental neoplasms. *Surgery* **91**, 452–458.

Boddie, A.W., Wright, K., Stephens, L.C., Yamanashi, W.S., Frazer, J., McBride, C.M., Wallace, S. and Martin, R.G. (1985). An animal model of occlusion-hyperthermia of the liver. *Invest. Radiol.* **20**, 159–165.

Brooks, D.J., Lammertsma, A.A., Beaney, R.P., Leenders, K.L., Buckingham, P.D., Marshall, J. and Jones, T. (1984). Measurement of regional cerebral pH in human subjects using continuous inhalation of $^{11}CO_2$ and positron emission tomography. *J. Cereb. Blood Flow Metab.* **4**, 458–465.

Burdon, R.H. (1984). Heat shock proteins. In *Hyperthermic Oncology*, Vol 2, pp 223–230. Overgaard, J. (Ed.), Taylor & Francis, London and Philadelphia.

Bull, J.M. (1983). Systemic hyperthermia background and principles. In *Hyperthermia in cancer therapy*, pp 401–405. Storm, F.K. (Ed.), G.K. Hall Medical Publishers, Boston.

Bull, J.M. (1984). A review of systemic hyperthermia. *Front. Radiat. Ther. Oncol.* **18**, 171–176.

Calderwood, S.K. and Dickson, J.A. (1983). pH and tumor response to hyperthermia. *Adv. Radiat. Biol.* **10**, 135–186.

Carr, K.E., Hume, S.P., Marigold, J.C.L. and Michalowski, A. (1982). Scanning and transmission electron microscopy of the damage to small intestinal mucosa following X-irradiation or hyperthermia. *Scan. Electron Microsc.* 1982 part I, 393–402.

Carr, K.E., Hume, S.P., Kamel, H.M.H., Marigold, J.C.L. and Michalowski, A. (1986). Development of villus damage in mouse small intestine after local hyperthermia compared with irradiation. *Int. J. Hyperthermia*. In press.

Crile, G. Jr. (1961). Heat as an adjunct to the treatment of cancer. Experimental studies. *Cleveland Clin. Quart.* **28**, 75–89.

Crile, G. Jr. (1963). The effects of heat and radiation on cancers implanted on the feet of mice. *Cancer Res.* **23**, 372–380.

Cronau, L.H., Bourke, D.L. and Bull, J.M. (1984). General anesthesia for whole-body hyperthermia. *Cancer Res. (Suppl.)* **44**, 4873s–4877s.

Cross, R.A. and Dewey, W.C. (1983). Mechanism of heat sensitization of G and S-phase cells by procaine-HC1. In *Tumour Biology and Therapy. Proceedings of the Seventh International Congress of Radiation Research*, pp D6–05. Broerse, J.J., Barendsen, G.W., Kal, H.B. and van der Kogel, A.J. (Eds), Martinus Nijhoff, Amsterdam.

Denekamp, J (1975). Changes in the rate of proliferation in normal tissues after irradiation. In *Radiation Research: Biomedical, Chemical and Physical Perspectives*, pp 810–825. Nygaaard, O.F., Adler, H.I. and Sinclair, W.K. (Eds.). Academic Press, New York and San Francisco.

Dewey, W.C., Westra, A., Miller, H.H. and Nagasawa, H. (1971). Heat-induced lethality and chromosomal damage in synchronized Chinese hamster cells treated with 5-bromodeoxyuridine. *Int. J. Radiat. Biol.* **20**, 505–520.

Dewey, W.C., Hopwood, L.E., Sapareto, S.A. and Gerweck, L.E. (1977). Cellular responses to combinations of hyperthermia and radiation. *Radiology* **123**, 463–474.

Dewey, W.C., Sapareto, S.A. and Betten, D.A., (1978). Hyperthermic radiosensitization of synchronous Chinese hamster cells: relationship between lethality and chromosomal aberrations. *Radiat. Res.* **76**, 48–59.

Dewhirst, M.W., Sim, D.A., Gross, J. and Kundrat, M.A. (1984). Effect of heating rate on tumour and normal tissue microcirculatory function. In *Hyperthermic Oncology 1984* Vol 1, pp 177–180. Overgaard, J. (Ed.), Taylor & Francis, London and Philadelphia.

Dickson, J.A. and Calderwood, S.K. (1980). Temperature range and selective sensitivity of tumours to hyperthermia: a critical review. In *Thermal Characteristics of Tumors: Applications in Detection and Treatment*, pp 180–205. Jain, R.K. and Gullino, P.M. (Eds.), Ann. N.Y. Acad. Sci., 335.

Dickson, J.A. and Calderwood, S.K. (1983). Thermosensitivity of neoplastic tissue *in vivo*. In *Hyperthermia in Cancer Therapy*, pp 63–140. Storm, F.K. and Hall, G.K. (Eds.), Boston, MA.

Dickson, J.A. and Ellis, H.A. (1974). Stimulation of tumour cell dissemination by raised temperatures (42 °C) in rats with transplanted Yoshida tumours. *Nature* **248**, 354–358.

Douglas, M.A., Parks, L.C. and Bebin, J. (1981). Sudden myelopathy secondary to therapeutic total-body hyperthermia after spinal-cord irradiation. *New Engl. J. Med.* **10**, 583–585.

Dudar, T.E. and Jain, R.K. (1984). Differential response of normal and tumor microcirculation to hyperthermia. *Cancer Res* **44**, 605–612.

Eddy, (1980). Alterations in tumor microvasculature during hyperthermia. *Radiology* **137**, 515–521.

Elkon, D., Fechner, R.E., Homzie, M.J., Baker, D.G. and Constable, W.C. (1980). Response of mouse kidney to hyperthermia: pathology and temperature-dependence of injury. *Arch. Pathol. Lab. Med* **104**, 153–158.

Endrich, B. and Hammersen, F. (1984). Hyperthermia-induced changes in capillary ultrastructure. *Int. J. Microcirc. Clin. Exp.* **3**, 498.

Endrich, B., Zweifach, B.W., Reinhold, H.S. and Intaglietta, M. (1979). Quantitative studies of microcirculatory function in malignant tissue: influence of temperature on microvascular hemodynamics during the early growth of the BA 1112 rat sarcoma. *Int. J. Radiat. Oncol. Biol. Phys.* **5**, 2021–2030.

Endrich, B., Voges, J. and Lehmann, A. (1984). The microcirculation of the amelanotic melanoma A–MEL–3 during hyperthermia. In *Hyperthermic Oncology 1984*, Vol 1, pp 137–140. Overgaard, J. (Ed.), Taylor & Francis, London and Philadelphia.

Fajardo, L.F. (1984). Pathological effects of hyperthermia in normal tissues. *Cancer Res. (suppl.)* **44**, 4826s–4835s.

Fajardo, L.F., Schreiber, A.B., Kelly, N.I. and Hahn, G.M. (1985). Thermal sensitivity of endothelial cells. *Radiat. Res.* **103**, 276–285.

Falk, P. (1983). The effect of elevated temperatures on the vasculature of mouse jejunum. *Br. J. Radiol.* **56**, 41–49.

Field, S.B. and Morris, C.C. (1983). The relationship between heating time and temperature: its relevance to clinical hyperthermia. *Radiother. Oncol.* **1**, 179–186.

Field, S.B. and Morris, C.C. (1984). Application of the relationship between heating time and temperature for use as a measure of thermal dose. In *Hyperthermic Oncology 1984*, Vol 1, pp 183–186. Overgaard, J. (Ed.), Taylor & Francis, London and Philadelphia.

Field, S.B. and Morris, C.C. (1985). Experimental studies of thermotolerance *in vivo*. I. The baby rat tail model. *Int. J. Hyperthermia* **1**, 235–246.

Gerner, E.W. and Schneider, M.J. (1975). Induced thermal resistance in HeLa cells. *Nature* **256**, 500–502.

Gerner, E.W., Boone, R., Connor, W.G., Hicks, J.A. and Boone, M.L.M. (1976). A transient thermotolerant survival response produced by single thermal doses in HeLa cells. *Cancer Res.* **36**, 1035–1040.

Gerweck, L.E. (1977). Modification of cell lethality at elevated temperatures. The pH effect. *Radiat. Res.* **70**, 224–235.

Gerweck, L.E. and Delaney, T.F. (1984). Persistence of thermotolerance in slowly proliferating plateau-phase cells. *Radiat. Res.* **97**, 365–372.

Gerweck, L.E., Dahiberg, W.K. and Greco, B. (1983a). Effect of pH on single or fractionated heat treatments at 42–45 °C. *Cancer Res.* **43**, 1163–1167.

Gerweck, L.E., Majima, H. and Delaney, T.F. (1983b). Variability in the kinetics of thermotolerance decay. In *Tumour Biology and Therapy: Proceedings of the Seventh International Congress of Radiation Research*, Amsterdam 1983. Broerse, J.J., Barendsen, G.W., Kal, H.B. and van der Kogel, A.J. (Eds.), Martinus Nijhoff, Amsterdam.

Goldacre, R. and Sylven, B. (1962). On the access of blood-born dyes to various tumour regions. *Br. J. Cancer* **16**, 306–322.

Goldin, E.M. and Leeper, D.B. (1981a). The effect of low pH on thermotolerance induction using fractionated 45 °C hyperthermia. *Radiat. Res.* **85**, 472–479.

Goldin, E.M. and Leeper, D.B. (1981b). The effect of reduced pH on the induction of thermotolerance. *Radiology* **141**, 505–508.

Gullino, P.M. (1980). Influence of blood supply on thermal properties and metabolism of mammary carcinomas. In *Thermal Characteristics of Tumors: Applications in Detection and Treatment*, pp 1–18. Jain, R.K. and Gullino, P.M. (Eds.), Ann. N.Y. Acad. Sci., Vol 335.

Gullino, P.M., Pon-Nyong, Y. and Grantham, F.H. (1978). Relationship between temperature and blood supply or consumption of oxygen and glucose by rat mammary carcinomas. *J. Natl. Cancer Inst.* **60**, 835–847.

Gullino, P.M., Jain, R. and Grantham, F. (1982). Temperature gradients and local perfusion in a mammary carcinoma. *J. Natl. Cancer Inst.* **68**, 519-533.

Hahn, G.M. (1974). Metabolic aspects of the role of hyperthermia in mammalian cell inactivation and their possible relevance to cancer treatment. *Cancer Res.* **34**, 3117–3123.

Hahn, G.M. (1982). In *Hyperthermia and Cancer*, pp 41–53. Plenum Press, New York and London.

Hand, J.W. (1986). Heat delivery and thermometry in clinical hyperthermia. In *Hyperthermia and Therapy of Malignant Tumours*. Streffer, C. (Ed.), Springer-Verlag. In press.

Hand, J.W., Walker, H., Hornsey, S. and Field, S.B. (1979). Effects of hyperthermia on the mouse testis and its response to X-rays, as assayed by weight loss. *Int. J. Radiat. Biol.* **35**, 521–528.

Harisiadis, L., Sung, D. and Hall, E.J. (1977). Thermal tolerance and repair of thermal damage by cultured cells. *Radiology* **123**, 505–509.

Henle, K.J. (1980). Sensitization to hyperthermia below 43 °C induced in Chinese hamster ovary cells by step-down heating. *J. Natl. Cancer Inst.* **64**, 1479–1483.

Henle, K.J. (1982). Thermotolerance in the mouse jejunum. *J. Natl. Cancer Inst.* **68**, 1033–1036.

Henle, K.J. and Dethlefsen, L.A. (1978). Heat fractionation and thermotolerance: a review. *Cancer Res.* **38**, 1843–1851.

Henle, K.J. and Dethlefsen, L.A. (1980). Time-temperature relationships for heat-induced killing of mammalian cells. *Ann. N.Y. Acad. Sci.* **335**, 234–252.

Henle, K.J. and Dethlefsen, L.A. (1982). Heat fractionation and step-down heating of murine mammary tumors in the foot. *Natl. Cancer Inst. Monogr.* **61**, 283–285.

Henle, K.J. and Leeper, D.B. (1976). Interaction of hyperthermia and radiation in CHO cells: recovery kinetics. *Radiat. Res.* **66**, 505–518.

Henle, K.J., Bitner, A.F. and Dethlefsen, L.A. (1979). Induction of thermotolerance by multiple heat fractions in Chinese hamster ovary cells. *Cancer Res.* **39**, 2486–2491.

Henriques, F.C. (1947). Studies of Thermal Injury. V. The predictability and the significance of thermally induced rate processes leading to irreversible epidermal injury. *Arch. Pathol.* **43**, 489–502.

Hill, S.A. and Denekamp, J. (1979). The response of six mouse tumours to combined heat and X-rays: implications for therapy. *Br. J. Radiol.* **52**, 209–218.

Hill, S.A. and Denekamp, J. (1982). Histology as a method for determining thermal gradients in heat tumours. *Br. J. Radiol.* **55**, 651–656.

Hill, S.A., Joiner, M., Bremner, J., Smith, K.A. and Denekamp, J. (1983). Factors influencing thermal sensitivity of a mouse tumour. In *Tumour biology and therapy. Proceedings of the Seventh International Congress of Radiation Research*, pp D6–19. Broerse, J.J., Barendsen, G.W., Kal, H.B. and van der Kogel, A.J. (Eds.), Martinus Nijhoff, Amsterdam.

Hill, S.A., Smith, K.A. and Denekamp, J. (1985). Thermal enhancement of stromal injury as assayed by the tumour bed effect. *Strahlentherapie* **161**, 535 (Abstract).

Hofman, P., Lagendijk, J.J.W. and Schipper, J. (1984). The combination of radiotherapy with hyperthermia in protocolized clinical studies. In *Hyperthermic Oncology 1984*, Vol 1, pp 379–382. Overgaard, J. (Ed.), Taylor & Francis, London and Philadelphia.

Honess, D.J. and Bleehen, N.M. (1985). Potentiation of melphalan by systemic hyperthermia in mice: therapeutic gain for mouse lung microtumours. *Int. J. Hyperthermia* **1**, 57–68.

Hume, S.P. and Field, S.B. (1978). Hyperthermic sensitization of mouse intestine to damage by X-rays: the effect of sequence and temporal separation of the two treatments. *Brit. J. Radiol.* **51**, 302–307.

Hume, S.P. and Marigold, J.C.L. (1980). Transient, heat-induced thermal resistance in the small intestine of mouse. *Radiat. Res.* **82**, 526–535.

Hume, S.P. and Marigold, J.C.L. (1982). Increased hyperthermal response of previously irradiated mouse intestine. *Br. J. Radiol.* **55**, 438–443.

Hume, S.P. and Marigold, J.C.L. (1984). The development of thermotolerance in hyperthermal injury to the villus compartment of mouse small intestine. *Int. J. Radiat. Biol.* **45**, 439–447.

Hume, S.P. and Marigold, J.C.L. (1985). Time-temperature relationships for hyperthermal radiosensitisation in mouse intestine: Influence of thermotolerance. *Radiother. Oncol.* **3**, 165–171.

Hume, S.P. and Marigold, J.C.L. (1986). Thermal enhancement of radiation damage in previously irradiated mouse intestine. *Br. J. Radiol.* **59**, 53–59.

Hume, S.P. and Myers, R. (1984). An unexpected effect of hyperthermia on the expression of X-ray damage in mouse skin. *Radiat. Res.* **97**, 186–199.

Hume, S.P. and Myers, R. (1985). Acute and late damage in mouse tail following hyperthermia and X-irradiation. *Int. J. Hyperthermia* **1**, 349–357.

Hume, S.P., Rogers, M.A. and Field, S.B. (1978a). Two qualitatively different effects of hyperthermia on acid phosphatase staining in mouse spleen, dependent on the severity of the treatment. *Int. J. Radiat. Biol.* **34**, 401–409.

Hume, S.P., Rogers, M.A. and Field, S.B. (1978b). Heat-induced thermal resistance and its relationship to lysosomal response. *Int. J. Radiat. Biol.* **34**, 503–511.

Hume, S.P., Marigold, J.C.L. and Field, S.B. (1979a). The effect of local hyperthermia on the small intestine of the mouse. *Br. J. Radiol.* **52**, 657–662.

Hume, S.P., Robinson, J.E. and Hand, J.W. (1979b). The influence of blood flow on temperature distribution in the exteriorized mouse intestine during treatment by hyperthermia. *Br. J. Radiol.* **52**, 219–222.

Hume, S.P., Marigold, J.C.L. and Michalowski, A. (1983). The effect of local hyperthermia on nonproliferative, compared with proliferative, epithelial cells of the mouse intestinal mucosa. *Radiat. Res.* **94**, 252–262.

Jain, R.K. and Ward-Hartley, K. (1984). Tumor blood flow-characterization, modifications and role in hyperthermia. *IEEE Transactions of Sonics and Ultrasonics*, Vol SU–31, No. 5, 504–526.

Jain, R., Grantham, F. and Gullino, P. (1979). Blood flow and heat transfer to Walker 256 mammary carcinoma. *J. Natl. Cancer Inst.* **62**, 927–933.

Jansen, W. (1980). Combination of hyperthermia and radiation in the treatment of experimental tumours in mice. *PhD Thesis*, University of Amsterdam, The Netherlands.

Joshi, D.S. and Jung, H. (1979). Thermotolerance and sensitization induced in CHO cells by fractionated hyperthermic treatments at 38–45 °C. *Eur. J. Cancer* **15**, 345–350.

Jung, H. (1982). Interaction of thermotolerance and thermosensitization induced in CHO cells by combined hyperthermic treatments at 40 and 43 °C. *Radiat. Res.* **91**, 433–446.

Jung, H. (1983). Modification of thermal response by fractionation of hyperthermia. *Strahlentherapie* **159**, 67–72.

Jung, H. and Kolling, H. (1980). Induction of thermotolerance and sensitization in CHO cells by combined hyperthermic treatments at 40 and 43 °C. *Eur. J. Cancer* **16**, 1523–1528.

Kamel, H.M.H., Carr, K.E., Hume, S.P. and Marigold, J.C.L. (1985). Structural changes in mouse small intestinal villi following lower body hyperthermia. *Scan. Electron Micros.*, Part II, 849–858.

Kamura, T., Nielsen, O.S., Overgaard, J. and Andersen, A.H. (1982). Development of thermotolerance during fractionated hyperthermia in a solid tumor *in vivo*. *Cancer Res.* **42**, 1744–1748.

Kang, M.S., Song, C.W. and Levitt, S.H. (1980). Role of vascular function in response of tumors *in vivo* to hyperthermia. *Cancer Res.* **40**, 1130–1135.

Kapp, D.S. and Lord, P.F. (1983). Thermal tolerance to whole-body hyperthermia. *Int. J. Radiat. Oncol. Biol. Phys.* **9**, 917–921.

Karino, T., Koga, S., Meata, M., Hamazoe, R., Yumane, T. and Oda, M. (1984). Experimental and clinical studies on effects of hyperthermia on tumor blood flow. In *Hyperthermic Oncology 1984*, Vol 1, pp 173–176. Overgaard, J. (Ed.), Taylor & Francis, London and Philadelphia.

Ladell, W.S.S. (1964). Terrestrial animals in humid heat: man. In *Handbook of Physiology, Section 4: Adaptation to the Environment*, pp 625–659. Dill, D.B. (Ed.), Williams & Wilkins, Baltimore.

Law, M.P. (1979). Induced thermal resistance in the mouse ear: the relationship between heating time and temperature. *Int. J. Radiat. Biol.* **35**, 481–485.

Law, M.P. (1981). The induction of thermal resistance in the ear of the mouse by heating at temperatures ranging from 41.5 to 45.5 °C. *Radiat. Res.* **85**, 126–134.

Law, M.P. (1985). Thermotolerance induced in the mouse ear by fractionated hyperthermia depends on the interval between fractions. *Strahlentherapie* **161**, 541 (abstract).

Law, M.P. and Ahier, R.G. (1982). Long-term thermal sensitivity of previously irradiated skin. *Br. J. Radiol.* **55**, 913–915.

Law, M.P., Ahier, R.G. and Field, S.B. (1977). The response of mouse skin to combined hyperthermia and X-rays. *Int. J. Radiat. Biol.* **32**, 153–163.

Law, M.P., Ahier, R.G. and Field, S.B. (1978). The response of the mouse ear to heat applied alone or combined with X-rays. *Br. J. Radiol.* **51**, 132–138.

Law, M.P., Coultas, P.G. and Field, S.B. (1979). Induced thermal resistance in the mouse ear. *Br. J. Radiol.* **52**, 308–314.

Law, M.P., Ahier, R.G., Somaia, S. and Field, S.B. (1984a). The induction of thermotolerance in the ear of the mouse by fractionated hyperthermia. *Int. J. Radiat. Oncol. Biol. Phys.* **10**, 865–873.

Law, M.P., Ahier, R.G., Somaia, S. and Field, S.B. (1984b). Does thermotolerance increase during fractionated hyperthermia? In *Hyperthermic Oncology 1984*, Vol 1, pp 195–198. Overgaard, J. (Ed.), Taylor & Francis, London and Philadelphia.

Law, M.P., Ahier, R.G. and Somaia, S. (1984c). A long-term effect of X-rays on thermal sensitivity of the mouse ear. *Br. J. Radiol.* **57**, 729–731.

Law, M.P., Ahier, R.G. and Somaia, S. (1985a). Thermal enhancement of radiation damage in previously irradiated skin. *Br. J. Radiol.* **58**, 161–167.

Law, M.P., Ahier, R.G., Somaia, S. and Field, S.B. (1985b). Thermotolerance after fractionated hyperthermia: assessment of cell survival by response to X-rays. *Int. J. Hyperthermia.* **1**,359–369.

Lee, D.H.K. (1964). Terrestrial animals in dry heat: man in the dessert. In *Handbook of Physiology, Section 4: Adaptation to the Environment*, pp 551–582. Dill, D.B. (Ed.), Williams and Wilkins, Baltimore.

Leeper, D.B. (1984). Molecular and cellular mechanisms of hyperthermia alone or combined with other modalities. In *Hyperthermic Oncology 1984*, Vol 2, pp 9–40. Overgaard, J. (Ed.), Taylor & Francis, London and Philadelphia.

Li, G.C. and Hahn, G.M. (1984). Mechanisms of thermotolerance. In *Hyperthermic Oncology 1984*, Vol 2, pp 231–234. Overgaard, J. (Ed.), Taylor & Francis, London and Philadelphia.

Lindegaard, J.C. and Overgaard, J. (1985). Effect of step-down heating on the interaction between heat and radiation in a C3H mammary carcinoma *in vivo*. *Strahlentherapie*, **161**, 542 (Abstract).

Lokshina, A.M., Song, C.W., Rhee, J.G. and Levitt, S.H. (1985). Effect of fractionated heating on the blood flow in normal tissues. *Int. J. Hyperthermia* **1**, 117–129.

Ludgate, C.M., Webber, R.G., Pettigrew, R.T. and Smith, A.N. (1976). Coagulation defects following whole-body hyperthermia in the treatment of disseminated cancer: a limiting factor in treatment. *Clin. Oncol.* **2**, 219–225.

Maher, J., Urano, M., Rice, L. and Suit, H.D. (1981). Thermal resistance in a spontaneous murine tumour. *Br. J. Radiol.* **54**, 1086–1090.

Marigold, J.C.L., Hume, S.P. and Hand, J.W. (1985). Investigation of thermotolerance in mouse testis. *Int. J. Radiat. Biol.* **48**, 589–595.

Marmor, J.B. and Hahn, G.M. (1978). Ultrasound heating in previously irradiated sites. *Int. J. Radiat. Oncol. Biol. Phys.* **4**, 1029–1032.

Marmor, J.B., Pounds, D., Postic, T.B. and Hahn, G.M. (1979). Treatment of superficial human neoplasms by local hyperthermia induced by ultrasound. *Cancer* **43**, 188–197.

Martinez, A.A., Meshorer, A., Meyer, J.L., Hahn, G.M., Fajardo, L.F. and Prionas, S.D. (1983). Thermal sensitivity and thermotolerance in normal porcine tissues. *Cancer Res.* **43**, 2072–2075.

Milligan, A.J. (1984). Whole-body hyperthermia induction techniques. *Cancer Res. (Suppl.)*, **44**, 4869s–4872s.

Milligan, A.J. and Panjehpour, M.S. (1985). Canine normal and tumor tissue estimated blood flow during fractionated hyperthermia. *Int. J. Radiat. Oncol. Biol. Phys.* **11**, 1679–1684.

Milligan, A.J., Metz, J.A. and Leeper, D.B. (1984). Effect of intestinal hyperthermia in the Chinese hamster. *Int. J. Radiat. Oncol. Biol. Phys.* **10**, 259–263.

Miyakoshi, J., Hiraoka, M., Takahashi, M., Kano, E., Abe, M. and Heki, S–I. (1983). Skin responses to step-up and step-down heating in C_3H mice. *Int. J. Radiat. Oncol. Biol. Phys.* **9**, 1527–1532.

Mooibroek, J., Zywietz, F., Dikomey, E. and Jung, H. (1984). Thermotolerance kinetics and growth pattern changes in an experimental rat tumour (R1H) after hyperthermia. In *Hyperthermic Oncology 1984*, Vol 1, pp 215–218. Overgaard, J. (Ed.), Taylor & Francis, London and Philadelphia.

Moritz, A.R. and Henriques, F.C. (1947). Studies of thermal injury II. The relative importance of time and surface temperature in the causation of cutaneous burns. *Am. J. Path.* **23**, 695–720.

Morris, C.C. and Field, S.B. (1979). Influence of hyperthermia on the oxygen enhancement ratio for X-rays, measured *in vivo*. *Br. J. Cancer* **40**, 878–882.

Morris, C.C. and Field, S.B. (1985). The relationship between heating time and temperature for rat tail necrosis with and without occlusion of the blood supply. *Int. J. Radiat. Biol.* **47**, 41–48.

Morris, C.C., Myers, R. and Field, S.B. (1977). The response of the rat tail to hyperthermia. *Br. J. Radiol.* **50**, 576–58.

Murata, T., Hasegawa, T. and Tanaka, Y. (1984). The effectiveness of arterial therapeutic blockage combined with hyperthermia in the treatment of tumors. In *Hyperthermic Oncology 1984*, Vol 1, pp 141–144. Overgaard, J. (Ed.), Taylor & Francis, London and Philadelphia.

Myers, R. (1983). A study of the combined effects of hyperthermia and irradiation on growing cartilage. *PhD thesis*, University of London, U.K.

Myers, R., Rogers, M.A. and Hume, S.P. (1981). Histochemical evidence for radiation enhancement of lysosomal response following hyperthermia of tail cartilage in baby rat. *Radiat. Res.* **87**, 329–340.

Nakayama, T. and Nakamura, W. (1984). Platelet aggregation induced in mice by whole-body hyperthermia. *Radiat. Res.* **98**, 583–590.

Neville, A.J., Robins, H.I., Martin, P., Gilchrist, K.W., Dennis, W.H. and Steeves, R.A. (1984). Effect of whole-body hyperthermia and BCNU on the development of radiation myelitis in the rat. *Int. J. Radiat. Biol.* **46**, 417–420.
Ng, T.C., Evanochko, W.T., Hiramoto, R.N., Ghanta, V.K., Lilly, M.B., Lawson,, A.J., Corbett, T.H., Durant, J.R. and Glickson, J.D. (1982). ^{31}P NMR Spectroscopy of *in vivo* tumors. *J. Magn. Reson.* **49**, 271–286.
Nielsen, O.S. and Overgaard, J. (1979). Effect of extracellular pH on thermotolerance and recovery of hyperthermic damage *in vitro*. *Cancer Res.* **39**, 2772–2778.
Nielsen, O.S. and Overgaard, J. (1982a). Importance of preheating temperature and time for the induction of thermotolerance in a solid tumour *in vivo*. *Br. J. Cancer.* **46**, 894–903.
Nielsen, O.S. and Overgaard, J. (1982b). Influence of time and temperature on the kinetics of thermotolerance in L1A2 cells *in vitro*. *Cancer Res.* **42**, 4190–4196.
Nielsen, O.S. and Overgaard, J. (1985). Studies on fractionated hyperthermia in L1A2 tumour cells *in vitro*: response to multiple equal heat fractions. *Int. J. Hyperthermia* **1**, 193–203.
Nielsen, O.S., Henle, K.J. and Overgaard, J. (1982). Arrhenius analysis of survival curves from thermotolerant and step-down heated L1A2 cells *in vitro*. *Radiat. Res.* **91**, 468–482.
Nielsen, O.S., Overgaard, J. and Kamura, T. (1983). Influence of thermotolerance on the interaction between hyperthermia and radiation in a solid tumour *in vivo*. *Br. J. Radiol.* **56**, 267–273.
Okumura, Y. and Reinhold, H.S. (1978). Heat sensitivity of rat skin. *Eur. J. Cancer* **14**, 1161–1166.
Overgaard, J. (1976a). Ultrastructure of a murine mammary carcinoma exposed to hyperthermia *in vivo*. *Cancer Res.* **36**, 983–987.
Overgaard, J. (1976b). Influence of extracellular pH on the viability and morphology of tumor cells exposed to hyperthermia. *J. Natl. Cancer Inst.* **8**, 1243–1250.
Overgaard, J. and Bichel, P. (1977). The influence of hypoxia and acidity on the hyperthermic response of malignant cells *in vitro*. *Radiology* **123**, 511–514.
Overgaard, J. and Nielsen, O.S. (1983). The importance of thermotolerance for the clinical treatment with hyperthermia. *Radiother. Oncol.* **1**, 167–178.
Overgaard, J. and Nielsen, O.S. (1984). Influence of thermotolerance on the effect of multifractionated hyperthermia in a C3H mammary carcinoma *in vivo*. In *Hyperthermic Oncology 1984*, Vol 1, pp 211–214. Overgaard, J. (Ed.), Taylor & Francis, London and Philadelphia.
Overgaard, K. and Overgaard, J. (1972). Investigations on the possibility of thermic tumour therapy, I. Short-wave treatment of a transplanted isologous mouse mammary carcinoma. *Eur. J. Cancer* **8**, 65–78.
Overgaard, J. and Suit, H.D. (1979). Time-temperature relationship in hyperthermic treatment of malignant and normal tissue *in vivo*. *Cancer Res.* **39**, 3248–3253.
Overgaard, J., Suit, H.D. and Walker, A.M. (1980). Multifractionated hyperthermia treatment of malignant and normal tissue *in vivo*. *Cancer Res.* **40**, 2045–2050.
Palzer, R.J. and Heidelberger, C. (1973). Studies on the quantitative biology of hyperthermic killing of HeLa cells. *Cancer Res.* **33**, 415–421.
Parks, L.C. and Smith, G.V. (1983). Systemic hyperthermia by extracorporeal induction: techniques and results. In *Hyperthermia in cancer therapy*, pp 407–446. Storm, F.K. (Ed.). G.K. Hall Medical Publishers, Boston.
Parks, L.C., Minaberry, D., Smith, D.P. and Neely, W.A. (1979). Treatment of far advanced bronchogenic carcinoma by extracorporeally induced systemic hyperthermia. *Thorac. Cardiovasc. Surg.* **78**, 883–892.
Peck, J.W. and Gibbs, F.A. (1983). Capillary blood flow in murine tumors, feet and intestines during localized hyperthermia. *Radiat. Res.* **96**, 65–81.
Pettigrew, R.T., Galt, J.M., Ludgate, C.M., Horn, D.B. and Smith, A.N. (1974). Circulatory and biochemical effects of whole-body hyperthermia *Br. J. Surg.* **61**, 727–730.
Prionas, S.D., Taylor, M.A., Fajardo, L.F., Kelly, N.I., Nelsen, T.S. and Hahn, G.M. (1985). Thermal sensitivity to single and double heat treatments in normal canine liver. *Cancer Res.* **45**, 4791–4797.
Rappaport, D.S. and Song, C.W. (1983). Blood flow and intravascular volume of mammary adenocarcinoma 13726A and normal tissues of rat during and following hyperthermia. *Int. J. Radiat. Oncol. Biol. Phys.* **9**, 539–547.
Raaphorst, G.P., Romano, S.L., Mitchell, J.B., Bedford, J.S. and Dewey, W.C. (1979). Intrinsic differences in heat and or X-ray sensitivity of seven mammalian cell lines cultured and treated under identical conditions. *Cancer Res.* **39**, 396–401.
Reinhold, H.S. and Endrich, B. (1986). Tumour microcirculation as a target for hyperthermia; a review. *Int. J. Hyperthermia* **2** 111–137.

Reinhold, H.S. and van den Berg-Blok, A.E. (1983a). Hyperthermia-induced alteration in erythrocyte velocity in tumors. *Int. J. Microcirc. Clin. Exp.* **2**, 285–295.

Reinhold, H.S. and van den Berg-Blok, A.E. (1983b). The influence of a heat pulse on the thermally induced damage to tumour microcirculation. *Eur. J. Cancer. Clin. Oncol.* **19**, 221–225.

Reinhold, H.S., Blachiewicz, B. and van den Berg-Blok, A. (1978). Decrease in tumor microcirculation during hyperthermia. In *Cancer Therapy by Hyperthermia and Radiation*, pp 231–232. Streffer, C. (Ed.), Urban & Schwarzenberg, Baltimore and Munich.

Rhee, J.G. and Song, C.W. (1984). Thermosensitivity of bovine aortic endothelial cells in culture: *in vitro* clonogenicity study. In *Hyperthermic Oncology 1984*, Vol 1, pp 157–160. Overgaard, J. (Ed.), Taylor & Francis, London and Philadelphia.

Rice, L.C., Urano, M. and Maher, J. (1982). The kinetics of thermotolerance in the mouse foot. *Radiat. Res.* **89**, 291–297.

Robins, H.I., Dennis, W.H., Neville, A.J., Shecterie, L.M., Martin, P.A., Grossman, J., Davis, T.E., Neville, S.R., Gillis, W.K. and Rusy, B.F. (1985). A nontoxic system for 41.8 °C whole-body hyperthermia: results of a phase I study using a radiant heat device. *Cancer Res.* **45**, 3937–3944.

Robinson, J.E., McCready, W.A. and Slawson, R.G. (1978). Thermal sensitivity of mouse mammary tumors. In *Cancer Therapy by Hyperthermia and Radiation*. Streffer, C., van Beuningen, D., Dietzel, F., Rottinger, E., Robinson, J.E., Scherer, E., Seeber, S. and Trott, K–R (Eds), Urban & Schwarzenberg, Baltimore and Munich.

Rogers, M.A., Marigold, J.C.L. and Hume, S.P. (1983). The effect of retinol on the hyperthermal response of normal tissue *in vivo*. *Radiat. Res.* **95**, 165–174.

Rottenberg, D.A., Ginos, J.Z., Kearfott, K.J., Junck, L., Dhawan, V. and Jarden, J.O. (1985). *In vivo* measurement of brain tumor pH using [^{11}C] DMO and positron emission tomography. *Ann. Neurol.* **17**, 70–79.

Sapareto, S.A. and Dewey, W.C. (1984). Thermal dose determination in cancer therapy. *Int. J. Radiat. Oncol. Biol. Phys.* **10**, 787–800.

Shibata, H. and MacLean, L. (1966). Blood flow to tumors. *Prog. Clin. Cancer* **11**, 33–47.

Sminia, P., Lebesque, J.V., van Dijk, J.D.P. and Haveman, J. (1985). A microwave heating technique for the rat spinal cord; thermometry and first results on the acute effects of hyperthermia on cervical spinal cord. *Strahlentherapie* **161**, 550 (Abstract).

Song, C.W. (1978). Effect of hyperthermia on vascular functions of normal tissues and experimental tumors: brief communication. *J. Natl. Cancer Inst.* **60**, 711–713.

Song, C.W. (1984). Effect of local hyperthermia on blood flow and microenvironment: a review. *Cancer Res.* (suppl.) **44**, 4721s–4730s.

Song, C.W., Kang, M.S., Rhee, J.G. and Levitt, S.H. (1980). The effect of hyperthermia on vascular function, pH and cell survival. *Radiology* **137**, 795–803.

Spiro, I.J., Sapareto, S.A., Raaphorst, G.P. and Dewey, W.C. (1982). The effect of chronic and acute heat conditioning on the development of thermal tolerance. *Int. J. Radiat. Oncol. Biol. Phys.* **8**, 53–58.

Stewart, F. and Begg, A. (1983). Blood flow changes in transplanted mouse tumours and skin after mild hyperthermia. *Br. J. Radiol.* **56**, 477–482.

Storm, F.K., Harrison, W.H., Elliott, R.S. and Morton, D.L. (1979). Normal tissue and solid tumor effects of hyperthermia in animal models and clinical trials. *Cancer Res.* **39**, 2245–2251.

Straw, J., Hart, M., Klubes, P., Zaharks, D. and Dedrick, R. (1974). Distribution of anticancer agents in spontaneous animal tumors. I. Regional blood flow and methotrexate distribution in canine lymphosarcoma. *J. Natl. Cancer Inst.* **52**, 1327–1331.

Streffer, C. (1985). Metabolic changes during and after hyperthermia. *Int. J. Hyperthermia* **1**, 305–319.

Strohbehn, J.W. and Douple, E.B. (1984). Hyperthermia and cancer therapy: a review of biomedical engineering contributions and challenges. *IEEE Trans. Biomed. Eng.* **BME–31**, 779–787.

Tamulevicius, P. and Streffer, C. (1983). Does hyperthermia produce increased lysosomal enzyme activity? *Int. J. Radiat. Biol.* **43**, 321–327.

Thistlethwaite, A.J., Leeper, D.B., Moylan, D.J. and Nerlinger, R.E. (1985). pH distribution in human tumors. *Int. J. Radiat. Oncol. Biol. Phys.* **11**, 1647–1652.

Tsubouchi, S., Kano, E., Nishimoto, Y. and Nakamura, W. (1984). The effect of microwave-induced whole-body hyperthermia on the mucosa of mouse small intestine. *J. Radiat. Res.* **25**, 131–139.

Urano, M. and Kahn, J. (1983). The effect of step-down heating on murine normal and tumor tissues. *Radiat. Res.* **94**, 350–358.

Urano, M. and Kahn, J. (1985). Effect of 42 °C hyperthermia on murine normal and tumor tissues. *Cancer Res.* **45**, 2527–2532.

Urano, M., Rice, L., Kahn, J. and Sedlacek, R.S. (1980). Studies on fractionated hyperthermia in experimental animal systems I. The foot reaction after equal doses: heat resistance and repopulation. *Int. J. Radiat. Oncol. Biol. Phys.* **6**, 1519–1523.

Urano, M., Cochran, L. and Montoya, V. (1982). Studies on fractionated hyperthermia in experimental animal systems II. Response of murine tumors to two or more doses. *Int. J. Radiat. Oncol. Biol. Phys.* **8**, 227–233.

Urano, M., Yamashita, T., Suit, H.D. and Gerweck, L.E. (1984). Enhancement of thermal response of normal and malignant tissues by *Corynebacterium parvum*. *Cancer Res.* **44**, 2341–2347.

van Beuningen, D., Issa, M., Breipohl, W., Streffer, C. and Rauwolf, M. (1983). Light- and electron-microscopical investigations on the effect of hyperthermia on the small intestine of mice. *Strahlentherapie* **159**, 367 (Abstract).

van der Zee, J., van Rhoon, G.C., Wike-Hooley, J.L., Faithfull, N.S. and Reinhold, H.S. (1983). Whole-body hyperthermia in cancer therapy: a report of a phase I–III study. *Eur. J. Cancer Clin. Oncol.* **19**, 1189–1200.

Vaupel, P., Muller–Klieser, W., Otte, J., Manz, R. and Kallinowski, M.F. (1983). Blood flow, tissue oxygenation and pH-distribution in malignant tumors upon localized hyperthermia. *Strahlentherapie* **159**, No. 2, 73–81.

Werts, E.D. and Smith, K.M. (1984). Temporal response of murine bone marrow to local hyperthermia. *Int. J. Radiat. Oncol. Biol. Phys.* **10**, 2315–2321.

Westra, A. and Dewey, W.C. (1971). Variation in sensitivity to heat shock during the cell-cycle of Chinese hamster cells *in vitro*. *Int. J. Radiat. Biol.* **19**, 467–477.

Wheldon, T.E., Hingston, E.C. and Ledda, J.L. (1982). Hyperthermia response and thermotolerance capacity of an experimental rat tumour with occluded blood flow. *Eur. J. Cancer Clin. Oncol.* **18**, 1007–1015.

Wike-Hooley, J.L., van der Zee, J., van Rhoon, G.C., van den Berg, A.P. and Reinhold, H.S. (1984). Human tumour pH changes following hyperthermia and radiation therapy. *Eur. J. Cancer Clin. Oncol.* **20**, 619–623.

Wills, E.J., Findlay, J.M. and McManus, J.P.A. (1976). Effects of hyperthermia therapy on the liver. II. Morphological observations. *J. Clin. Pathol.* **29**, 1–10.

Wondergem, J. and Haveman, J. (1983). The response of previously irradiated mouse skin to heat alone or combined with irradiation: influence of thermotolerance. *Int. J. Radiat. Biol.* **44**, 539–552.

Wondergem, J. and Haveman, J. (1984). A study of the effects of prior heat treatment on the skin reaction of mouse feet after heat alone or combined with X-rays: influence of misonidazole. *Radiother. Oncol.* **2**, 159–170.

Wondergem, J. and Haveman, J. (1985). The sensitivity of normal stroma to heat alone or combined with irradiation measured by a tumour growth rate assay. *Strahlentherapie* **161**, 555 (Abstract).

Wynne, J.M., Mack, S., McRae, D., Pillay, S.P., Potts, J., Boffinger, C., Cowley, D.M. and Egerton, W.S. (1984). Portal vein perfusion of the isolated rat liver: Some markers of hyperthermic liver damage. *Aust. J. Exp. Biol. Med. Sci.* **62**, 73–80.

Hyperthermia and Oncology, Vol. 1, pp. 161–200 (1988)
Urano and Douple (Eds)
© 1988 VSP.

Chapter 6

Tumor response to hyperthermia

MUNEYASU URANO

Department of Radiation Medicine, Edwin L. Steele Laboratory, Massachusetts General Hospital, Harvard Medical School, Boston, MA 02114, USA

A. INTRODUCTION

Biological studies have demonstrated that hyperthermia can cause lethal damage to mammalian cells and enhance the effect of radiation and some chemotherapeutic agents. These effects on cultured mammalian cells have been observed in animal tumors as well. In this chapter the effect of hyperthermia alone on the tumor will be discussed together with the effect of some modifiers on thermal effect of the tumor.

Studies on the tumor response to heat have been performed from the end of the last century. Early studies were motivated by the recognition that tumor cells were intrinsically more sensitive to elevated temperatures than normal tissue cells (Schrek 1966; Cavaliere *et al.* 1967). However, current studies have disclosed that the intrinsic thermal sensitivity of tumor cells are identical to that of normal tissue cells (Kase and Hahn 1976; Symonds 1981), and that the differences in thermal sensitivity are due to environmental or pathophysiological differences between tumor and normal tissues (Gerweck 1985). Blood flow rate may differ between these tissues, as well as tissue pH, pO_2, pCO_2, and nutrient supply. Some pathophysiological features which are specific to the tumor tissue tend to increase thermal sensitivity of tumor cells.

A critical feature for the heat treatment of the tumor may be that the average tumor tissue pH is lower compared to the normal tissue pH. The low pH in the tumor may result from the presence of hypoxic tumor cells and the high glycolysis level of tumor cells. Uncontrolled tumor growth leads to a formation of unvascularized areas in the tumor. Tumor cells existing more than 200μm away from a capillary become anoxic and then become necrotic, since oxygen can perfuse only 200μm (Thomlinson and Gray 1955). Some tumor cells between the edge of anoxic or necrotic foci and the viable tumor cells are oxygen-deficient, but still maintain their metabolic activity. These cells undergo anaerobic glycolysis resulting in an accumulation of lactic acid which eventually reduces tumor tissue pH. Further metabolism of lactic acid may be prevented by the lack of oxygen.

Tumor cells utilize a substantial amount of glucose as compared to normal tissue cells, and metabolize it into lactic acid in the presence of normal respiration processes, i.e. in the presence of oxygen. This aerobic glycolysis is due to an anomaly in the integration of glycolytic sequence, and results in the accumulation of lactic acid in the tumor (Lehninger 1981). It is not realistic, however, to predict a uniform pH value throughout the tumor. The pH value may be normal or near normal in the area adjacent to capillaries, since lactic acid may be efficiently removed by the blood circulation. On the other hand, lactic acid may be accumulated in the avascularized area. Therefore, non-uniform pH distribution may be a feature of tumor tissue, as well as tumor cell heterogeneity in thermal sensitivity.

B. ASSAY METHODS OF TUMOR RESPONSE TO HYPERTHERMIA

Tumor response to hyperthermia depends on temperature and duration of treatment. The dose response relationship can be obtained by various assay methods as a function of treatment time at a given treatment temperature. The cell survival curve could be determined following hyperthermia given *in vivo*. Assay methods may include *in vivo* assays such as the TD_{50} assay* (Hewitt and Wilson 1959) and lung colony assay† (Hill and Bush 1969) or *in vitro* colony formation assay. Namely, animal tumors are treated at an elevated temperature induced using a water bath, radiofrequency currents, or other heating techniques, and removed immediately thereafter. Intact tumor tissues are minced and dispersed enzymatically or mechanically to prepare a single cell suspension. This suspension is serially diluted by medium and transplanted into subcutaneous tissues for TD_{50} assays, injected intraven-

---

*TD_{50} Assay: An assay method to determine the number of tumor cells to transplant a tumor into one-half of the transplanted sites. Recipient animals are randomly assigned into several cell dose groups. A single cell suspension is serially diluted and transplanted into the subcutaneous tissue. Each cell dose group may consist of 5 to 10 animals. Tumor takes are observed for 1 to 4 months, depending on the tumor growth rate. The TD_{50} value can be calculated by probit or logit analysis according to the tumor take rate in each dose group. The TD_{50} value is obtained for each treatment time and the surviving fraction is calculated by TD_{50} (control)/TD_{50} (test).

†Lung colony assay: A single cell suspension prepared from each treatment dose group is serially diluted into 1 to 3 cell dose groups and injected intravenously into recipient animals. Tumor cells lodge in the lung and each cell which has proliferative capability forms a colony in the lung. Approximately 5 animals are used in each cell dose group. Recipients are sacrificed 2 to 3 weeks thereafter and lungs are removed. After staining with Bouin's solution for at least overnight, colonies formed on the surface in each lobe are counted. Surviving fraction can be calculated as follows:

$$S.F. = \frac{\text{Number of colonies formed}}{\text{Number of treated cells injected} \times CFP}$$

where CFP stands for colony forming probability or:

$$CFP = \frac{\text{Number of colonies formed}}{\text{Number of untreated cells injected}}$$

ously for lung colony assays, or plated in petri dishes for *in vitro* colony assays. The cell survival can be determined on the basis of tumor take frequency or colony forming ability.

A pitfall in these assay methods may be in the early development of necrosis in some tumors following hyperthermia. Histological studies demonstrated some changes, such as swelling of the cytoplasm, thickening of the nuclear membrane and increased nuclear lucency 15 minutes after the beginning of hyperthermia at 44 °C. By 2.5 hours after the treatment, wide areas of necrosis, marked infiltration of segmented neutrophils, wide-spread hemorrhage and karyopyknosis of tumor cells are observed in histology sections (Fajardo *et al.* 1980).

The assay methods just mentioned must involve a cell dispersion technique, either enzymatic or mechanical. The calculation of surviving fractions is based on the assumption that all tumor cells dispersed are morphologically intact regardless of their proliferative capability. Early appearance of pyknotic tumor cells in some tumors may discredit this assumption and make the obtained survival fractions somewhat unreliable.

TCD_{50} assay, or the determination of the treatment time to yield a tumor cure in one-half of the treated animals (Suit *et al.* 1965; Urano *et al.* 1980) is another possible technique to examine the tumor response. We have investigated the TCD_{50} of four spontaneous, non-immunogenic murine tumors transplanted into the mouse foot. None of these tumors have been controlled without showing severe normal tissue damage, including partial to total loss of the tumor bed (the foot) induced within 5 weeks after heat treatment. This indicates that the TCD_{50} value may be strongly influenced by the tumor bed damage. Presumably, a small number of tumor cells which survived the heat treatment might be lost together with the tumor bed without manifesting a recurrence. The TCD_{50} values for immunogenic tumors can be obtained without severe normal tissue damage. Some data from these experiments are shown in the next section.

The TG (tumor growth) time assay is probably most commonly used for studying the tumor response to hyperthermia (Urano and Kahn 1987). The TG time is the time required for one-half of the treated tumors to reach a specified volume from the treatment day. This method measures tumor regrowth time. Although the TG time is not only related to thermal sensitivity of tumor cells, but also depends on the tumor cell kinetics, such as cell cycle time, growth fraction and cell loss factor. The tumor bed effect may influence the tumor regrowth rate after a large heat dose. An advantage of this method is that the dose response relationship between the treatment time and the TG time can be obtained. Unlike the cell survival curve, the slope of the dose response curve is not a direct measure of the thermal sensitivity of the tumor cell, although it is proportional to it. Accordingly, thermal sensitivity of various tumors cannot be evaluated by the slopes of this type of dose response relationship. Comparison of the slopes can only be made for the experiment using the *same* animal tumor system e.g., for the experiment comparing different treatment modalities, or for the study of thermal modifiers in an animal tumor system.

The TG time of some tumors linearly relates to the treatment time at an elevated temperature, while that of some other tumors exponentially relates to the treatment time. The slopes of these two types of dose response curves are by no means comparable. An example of each is shown in Figs. 1 and 2, respectively in the next section.

The TGD (tumor growth delay) time is the difference between the TG time (treated) and the TG time (no treatment). It can be obtained by subtracting the TG time (no treatment) from the TG time (treated) and the dose response relationship between the TGD time and the treatment time can also be obtained.

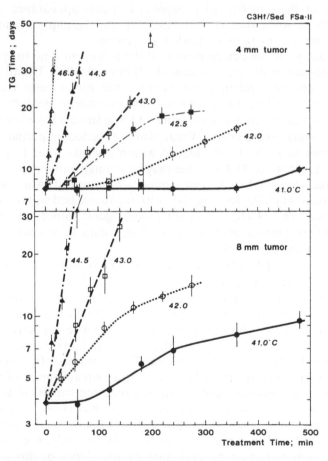

Figure 1: Dose response curves for the C3Hf/Sed mouse FSa–II tumors treated at various temperatures. Logarithm of the TG time to reach 1000 mm³ is plotted as a function of treatment time at each elevated temperature as indicated in the figure. Upper and lower panels are for 4 and 8mm tumors, respectively (from Urano *et al.* 1982 with permission).

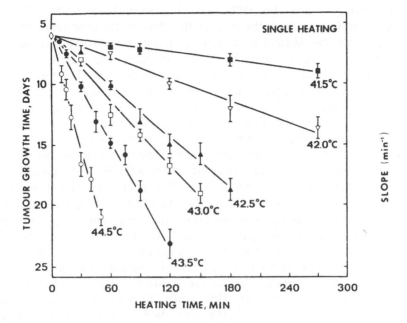

Figure 2: Dose response curves for a C3H mouse mammary carcinoma (200 mm³). The TG time is plotted as a function of treatment time. Treatment temperatures are indicated in the figure (from Neilsen and Overgaard 1982 with permission).

Water bath heating is often used for small animal experiments since the protruding parts of the body including the foot, tail and ear can be easily heated by immersing into the water bath. The foot is a frequent part of tumor transplantation because the water bath heating of the foot tumor allows relatively uniform temperature distribution.

C. EFFECT OF TREATMENT TEMPERATURE AND TUMOR SIZE ON THE TUMOR RESPONSE

1. Treatment temperature

Tumor response to hyperthermia depends on the treatment temperature and the duration of the treatment. It is modified by various environmental, pathophysiological factors, and by some chemical modifiers.

Top panel of Fig. 1 shows the TG time of 4mm FSa-II tumor (a murine spontaneous fibrosarcoma with an average diameter of 4mm) as a function of treatment time at various treatment temperatures, where the TG time means the time required for one-half of the treated tumors to reach 1000 mm³ from the treatment day (Urano *et al*. 1982). The dose response curves obtained at temperatures above 43.0 °C are exponential with or without an initial shoulder. However, the dose response curves below 43.0 °C are biphasic, indicating the development of resistance to continuous heating. Currently, this resistant tail is well-defined as the development of thermotolerance during hyperthermia (Urano 1986). It is clear from the figure that the higher the treatment temperature, the steeper the slope of the dose response curve. In other words, the thermal sensitivity of the tumor increases with increasing temperature. For the exponential portion of each dose response curve, an exponential regression line can be fitted between the TG time and the treatment time for each treatment temperature. The slope of the regression line does not directly reflect thermal sensitivity of the tumor cell and is not equivalent to the D_0 (the treatment time to reduce the survival from 1.0 to 1/e on the exponential portion of the cell survival curve), but it should be proportional to the D_0. Therefore, the slope which has a unit of min $^{-1}$ should be proportional to the rate constant for cell killing at a given temperature. The Arrhenius plot (see Chapter 3) can be constructed by plotting the slope as a function of $1/T$ where T is the absolute temperature (Fig. 3). The slope of the Arrhenius plot represents the activation energy for thermal cell killing and has a breaking point at 42.5 °C. Above this temperature, the activation energy is 145 kcal/M, which is identical to the activation energy for many cultured mammalian cells. This also indicates that a 1 °C increase in the treatment temperature halves the treatment time to obtain the same tumor response (e.g., in the TG time study, to obtain the same TG time). At temperatures below 42.5 °C, the activation energy is greater than 145 kcal/M, but it cannot be calculated from two data points. It is notable that the shoulder width or the threshold time of the dose response curve increases with increasing temperatures. The size of the shoulder does not influence the D_0 value, but it has a strong impact on the tumor therapy.

The lower panel of Fig. 1 demonstrates the dose response relationships of the 8 mm FSa-II tumor (tumors with an average diameter of 8 mm). In spite of this being the same FSa-II tumor, the slope of the dose response curve of the 8 mm tumor is steeper than that of the 4 mm tumor. This indicates that the 8 mm tumor is more sensitive to heat than the 4 mm tumor providing that the cell kinetics of 4 and 8 mm tumors and the tumor bed effect for two-sized tumors are identical following hyperthermia. The development of resistance is also observed during the treatment at 42.0 and 41.0 °C. The slope is plotted as a function of treatment time to construct the Arrhenius plot. Unlike the 4mm tumor, no breaking point is found in the temperature range between 41.0 and 45.5 °C (Fig. 3). The activation energy is identical to that of the 4 mm tumor. Unlike the 4 mm tumor, an increased shoulder width is observed only at 41.0 °C. The difference in thermal sensitivity between two different-sized tumors will be discussed later.

Nielsen and Overgaard (1982) investigated the dose response relationships at various temperatures in a mouse mammary carcinoma. A linear relationship was found between the TG time and the treatment time (Fig. 2). The slope is proportional to the D_0 as discussed above and can be plotted on the Arrhenius plot (Fig. 3). The activation energy is 150 kcal/M above 42.5 °C, which is in the range reported for the cultured mammalian cells. At temperatures below 42.5 °C, the activation energy increased to 360 kcal/M. It is notable that the slope of the mammary carcinoma cannot be directly compared with that of the FSa-II tumor since the latter is derived from exponential regression lines, while the former is from linear regression lines. It is also notable that these slopes do not represent absolute values of the rate constant. They are only proportional to it and, therefore, only allow the calculation of the activation energy.

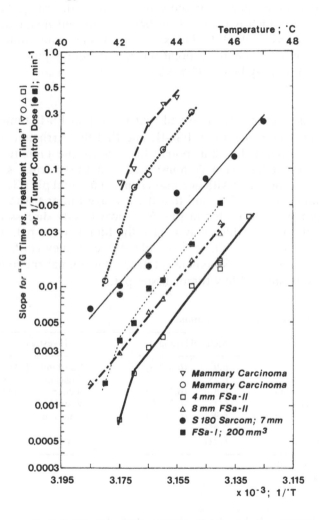

Figure 3: Arrhenius plots for various animal tumors. The slope of the dose response curve between TG time and treatment time (open symbols) or the reciprocal of TCD_{50} value (solid symbols) (both have a unit, min^{-1}) is plotted as a function of the reciprocal of T (absolute temperature). Activation energies are in the range between 130 and 155 kcal/mol.

In Fig. 3 data taken from another experiment are also shown which exhibit a breaking point at 43.0 °C (Robinson *et al.* 1978). Solid symbols in the figure are taken from two experiments investigating the tumor cure (Crile 1963; Overgaard and Suit 1979). Tumors are immunogenic and the tumor control is obtained without severe normal tissue damage. The tumor control dose depends on the D_0. Therefore, the reciprocal of the tumor control dose is proportional to $1/D_0$, or the rate constant. No breaking point is seen for SA180 sarcoma, while it is observed between 42.5 °C and 42.0 °C for the FSa-I tumor (methylcholanthrene-induced C3H mouse fibrosarcoma). As a summary, the activation energy for these tumors is in the range between 130 and 155 kcal/M, which is identical to that for cultured mammalian cells at temperatures above $\simeq$ 43.0 °C. Unlike cultured cells, some tumors or some sized tumors show no breaking point on the Arrhenius plot and retain the same activation energy below 42.5 °C.

2. Tumor size

In the previous section, we observed that the 8 mm FS-II tumo was more sensitive to heat than the 4 mm FSa-II, and that the Arrhenius plot for the 8 mm tumor showed no breaking point while that for the 4 mm tumor showed it at $\simeq$ 42.5 °C. It has been demonstrated that the thermal sensitivity of mammalian cells increases with decreasing environmental pH (Gerweck 1977; Overgaard and Bichel 1977), and that the average tumor tissue pH is lower than the normal tissue or blood pH. We have briefly discussed extensive glycolysis in the tumor which may reduce the tumor tissue pH. Table 1 lists reported data for murine tumor tissue pH. Wike-Hooley *et al.* (1984) summarized numerous pH measurements performed in animal and human tumors that consistently showed low tumor tissue pH.

Table 1.
The mean pH value in solid mouse tumors

Tumor	Mean pH (range)		Investigator	
MC-induced	6.9	(6.6–7.3)*	Naesland	(1953)
Adenocarcinoma	6.25		Tagashira	(1955)
Mammary carcinoma	6.8	(6.2–7.1)	Bicher	(1980)
Mammary carcinoma	6.75	(6.2–7.0)	Vaupel	(1982)
SCK mammary carcinoma	6.95	(6.6–7.3)	Rhee *et al.*	(1984)
FSa–II 4–6mm	7.05†		Rhee	
FSa–II 8–10mm	6.92†		Rhee	

* Calculated by author.
† Unpublished data.

In some tumors, it has been shown that the average tumor tissue pH decreases with increasing tumor size (Kahler and Moore 1962; Jahde *et al.* 1982; Jain *et al.* 1984) (Fig. 4). This finding may be reasonable in some tumors which show an increase in the size of hypoxic cell fraction and in the size of necrosis, and a decrease in the blood flow rate with increasing tumor size (Song *et al.* 1980). The size of hypoxic cell fraction of the 4 mm FSa-II tumor is $\simeq$ 4 per cent, while that of 8 mm FSa-II is $\simeq$ 40 per cent (Urano, unpublished data). In these situations, an average tumor pH is lower in the large tumor than in the small tumor and, therefore, the large tumor could be more sensitive to heat than the small tumor.

The pH distribution in the tumor tissue is not uniform, as discussed in the introduction. This indicates that the thermal sensitivity of tumor cells is not uniform since the thermal sensitivity depends on the environmental pH. This pH effect is greater at temperatures below 43.0 °C than at temperatures above 43.0 °C (Gerweck 1977). Greater pH effect below 43.0 °C may be attributed to the following two factors: (a) the activation energy for cells in normal pH conditions increases below 43.0 °C while the same activation energy as that above 43.0 °C is maintained below 43.0 °C for cells in the low pH; and (b) below 43.0 °C thermotolerance develops during the heat treatment in normal pH conditions, while no thermotolerance can be detected in cells in lower pH conditions. As a result, the difference in thermal sensitivity between cells in pH 7.4 and cells in 6.7 is greater at temperatures below 43.0 °C compared to temperatures above 43.0 °C. It might be ≃ 1.5 fold at temperatures above 43.0 °C, while this difference could be 3 times or greater at temperatures below 43.0 °C.

Thermal sensitivity of tumor cells varies greatly according to the environmental pH and nutritional conditions. For simplification, let us divide thermal sensitivity of all tumor cells into three categories; namely, resistant (cells in pH 7.4), intermediate (pH 7.0), and sensitive (pH 6.7) cells. The thermal

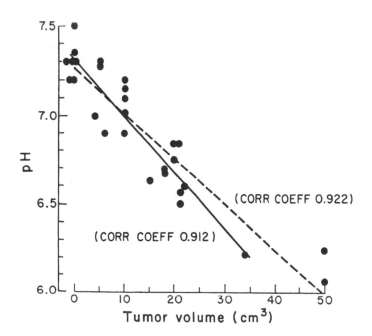

Figure 4: Tumor tissue pH as a function of tumor volume. A linear relationship is observed between the tissue pH and tumor size (dotted line). Solid line is for the tumor less than 35 cm³ (from Jain *et al.* 1984 with permission).

sensitivity of a tumor is the sum of the sensitivity of each tumor cell and, therefore, depends on the size of each cell fraction (Figs. 5 and 6).

Firstly, let us examine the thermal sensitivity of tumors treated at temperatures above 43.0 °C (Fig. 5, panels A–C). If 80 per cent of tumor cells are resistant to heat, the resistant cells dominate the whole population, and the overall survival curve of tumor cells is almost identical to that of the resistant

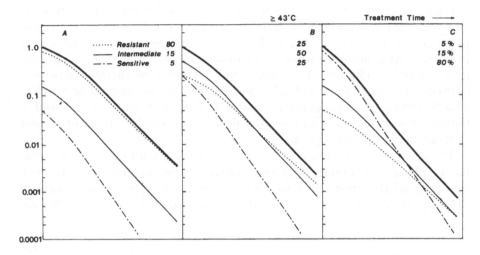

Figure 5: Schematic illustrations of tumor cell survival curves following hyperthermia given above 43.0 °C. It is assumed that tumors consist of three types of cells, i.e. resistant, intermediately sensitive, and sensitive cells to hyperthermia. Thin lines indicate survival curves of each type of tumor cells, and thick lines indicate the overall survival curves. The size of each cell fraction is indicated in each panel.

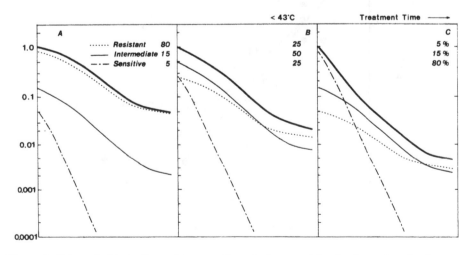

Figure 6: Schematic illustrations of tumor cell survival curves following hyperthermia given below 43 °C, or at the temperature below the breaking point on the Arrhenius plot. The same symbols as in Fig. 6 are used. Notably the pH effect is greater at this temperature and thermotolerance develops during the treatment in resistant cells (see text).

cells (panel A). On the other hand, if only 5 per cent of tumor cells are resistant and 80 per cent are sensitive, the overall survival curve may follow the sensitivity of the sensitive cells. With increasing treatment time, however, less resistant cells receive lethal damage compared to the sensitive cells, and the surviving fraction of resistant cells will become dominant over the sensitive cells. The overall survival curve may be slightly biphasic (panel C). However, the biphasic nature may be too slight to be distinguished from an exponential relationship. Panel B is an intermediate between panels A and C. The ratio of D_0 in panel A to that in panel C is approximately 1.25, if exponential regression lines are fitted to both overall survival curves. If any pH modifiers can reduce the pH in more than 99 per cent of the tumor cells to pH 6.0, the difference could be much greater. It is notable that the extrapolation number decreases with an increase in the size of the sensitive cell fraction. This evidence will be more noticeable at temperatures below 43.0 °C.

Secondly, let us examine the survival curves of cells treated at temperatures below 43.0 °C (Fig. 6). As mentioned above, the difference in the D_0 between pH 7.4 and pH 6.7 cells is much greater than the difference observed at temperatures above 43.0 °C. In addition, thermotolerance develops in the normal or near-normal pH cells, but it does not develop in the low pH cells. If 80 per cent of the tumor cells are resistant (panel A), the contribution of sensitive or intermediately sensitive cells to survival is trivial and the overall survival curve follows the survival curve of the resistant cells. If only 5 per cent of the tumor cells are resistant and sensitive cells dominate, the overall survival curve shows the initial sensitive portion. However, with increasing treatment time, the contribution of the resistant cell to the survival becomes dominant and the survival curve finally shows a resistant tail which is due to the development of thermotolerance in the resistant cells (panel C). This schematic illustration in panel C is comparable to the dose response curve of the 8 mm FSa-II tumor shown in Fig. 1 and explains the development of thermotolerance in the large tumor treated at 41 or 42 °C. Furthermore, as mentioned above, the extrapolation number of the survival curve decreases with an increase in the size of sensitive cell fraction, and becomes approximately 1.0 in panel C. This is also observed in the dose response curve of the 8mm tumors treated at 41.0 or 42.0 °C. Panel B demonstrates an intermediate overall survival curve between those in A and C. Unlike the overall survival curves following treatment at temperatures above 43.0 °C, the D_0 changes substantially depending on the size of the resistant and sensitive tumor cell fractions. The ratio of D_0 of the overall survival curve where resistant cells dominate (panel A) to D_0 of the overall survival curve where sensitive cells dominate (panel C) is $\simeq 1.7$, which can be compared with $\simeq 1.25$ obtained for the treatment above 43.0 °C. This large D_0 ratio can explain the observation that the same activation energy above 43.0 °C is retained at temperatures below 43.0 °C in the large tumor.

These schematic calculations explain the difference in thermal sensitivities between the small tumor in which the resistant cells dominate, and the large tumor in which the sensitive cells dominate. This also explains increasing difference in thermal sensitivity at temperatures below 43.0 °C between the small and large tumors. It is notable for clinical application of hyperthermia that thermotolerance develops at temperatures below 43.0 °C even in a tumor where only 5 per cent of tumor cells are resistant.

The non-uniform thermal sensitivity throughout the tumor is also observed histopathologically. Tumor cells adjacent to normal tissues and to blood vessels are most likely in normal pH conditions, and are presumably protected from thermal damage. Hill and Denekamp (1982) demonstrated histologically that intact tumor cells remain and regrow in these areas, while tumor cells in most other areas become necrotic following hyperthermia (Fig. 7, A and B).

In the above discussions, tumor tissue temperatures were assumed to be relatively uniform. However, uniform temperature distribution is difficult to obtain in clinical hyperthermia and in experimental animal tumors except tumors in the foot. The non-uniform temperature distribution results in the non-uniform thermal sensitivity since the thermal sensitivity depends on the treatment temperature. The presence of intact tumor cells around the capillary and near the normal tissue, shown in Fig. 7, may partially be attributed to the presence of cold spots where heat deposited may be quickly washed out by good blood circulation. Wallen et al. (1981) demonstrated that samples taken from rat tumors which had been heated by 2450 MHz microwave showed relative colony-forming efficacies which varied by factors > 100. Besides the pH and temperature distribution in the tumor, tumor cells are heterogeneous in thermal sensitivity. It depends on the cell cycle (Dewey et al. 1977), and non-cycling cells may have different thermal sensitivity.

D.   MODIFIERS OF THERMAL RESPONSE

Some chemical agents can enhance the response of mammalian cells and tissues to hyperthermia. Agents which can reduce extra- or intra-cellular pH and cellular energy status can sensitize thermal response.

Administration of glucose stimulates all pathways of glycolysis with a resultant accumulation of lactic acid which decreases the tissue pH. Forced breathing of air containing excess of CO_2 reduces blood and tissue pH by increasing pCO_2. Mechanical occlusion of blood vessels out of and into the tumor tissue may deprive oxygen and energy needed to sustain life of the cells and cause an accumulation of lactic acid. The resultant energy-deprived status and reduced tissue pH enhances the response of the tissue to hyperthermia. A potent male contraceptive, 5 thio-D-glucose, inhibits the anaerobic glycolytic pathway, leading to specific sensitization against hypoxic cells. An agent called gossypol is also a potential sensitizer by blocking glycolysis. Two agents, lonidamine and quercetin are known as inhibitors of lactate transport which, therefore, results in accumulation of the intracellular lactate. Two other types of chemicals which enhance thermal response are local anaesthetics and polyamines. The sensitizing effect of many chemical agents are not yet proven in vivo.

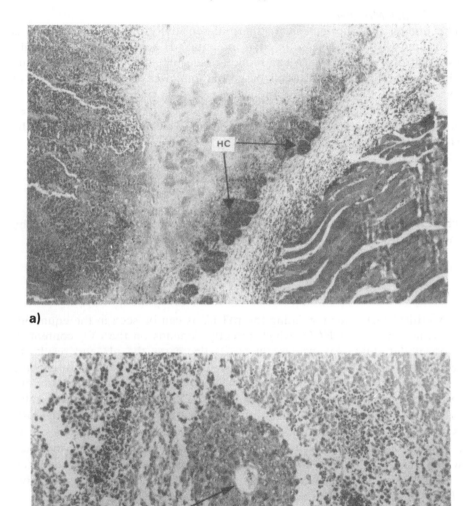

Figure 7: Histology sections of a rodent tumor treated at an elevated temperature. Morphologically intact cells are observed in the area adjacent to the normal tissues (muscle), or tumor periphery in photo A, and in the area surrounding blood vessels in photo B. Tumor cells in other areas are necrotic or pycnotic. These pictures indicate the inhomogeneity of the thermal sensitivity in the tumor (photographs were kindly provided by Dr S. Hill).

1. The pH Modifiers

The finding that the increase in thermal sensitivity of mammalian cells is associated with the decrease in environmental pH has stimulated the establishment of a method which can reduce tissue pH, specifically tumor tissue pH. Presumably, reduced tumor tissue pH enhances the tumor response to hyperthermia.

Acid-base balance in the normal animal body is physiologically regulated and the pH is stated in the general form which is referred to as the Henderson–Hasselbach equation, i.e.,

$$pH = pK + \log \frac{base}{acid}$$

where K is a constant. The pH of the extracellular fluid can be defined by the relation between sodium bicarbonate and carbonic acid, i.e.,

$$pH = 6.1 + \log \frac{NaHCO_3}{H_2CO_3}$$

A critical factor for regulating the pH is, as can be seen in the equation, the concentration of HCO_3 which directly depends on the CO_2 content in extracellular fluid. An increase in pCO_2 decreases the pH and results in respiratory acidoses, which is frequently observed in patients with chronic respiratory diseases. This indicates that acidoses can be experimentally obtained by giving excess of CO_2 in respiratory air. The excess CO_2 penetrates cell membrane freely and reduces intracellular pH. Another type of acidosis is caused by the accumulation of acids and is called metabolic acidosis. The metabolic acidosis occurs in patients with uncontrolled diabetes mellitus, which causes the accumulation of ketonic acid. Some patients with advanced cancer develop so-called lactate acidosis (Sculier et al. 1983) since the extensive glycolysis in malignant tumors produces a large amount of lactic acid.

a. Glucose. In an experimental situation, a glucose administration (hyperglycemia) specifically reduces tumor tissue pH (Voegtlin et al. 1936; Jahde and Rajewsky, 1982). Hyperglycemia stimulates the synthesis and release of insulin which facilitates all pathways of glycolysis. In the tumor tissue, both aerobic and anaerobic glycolysis, which is naturally more extensive in the tumor than in the normal tissue, may be facilitated. This process results in an extensive accumulation of lactic acid, which further reduces tumor tissue pH. Furthermore, hyperglycemia increases extracellular osmotic pressure (glucose is a major molecule which controls extracellular osmotic pressure) with resultant intracellular dehydration, and increases blood viscosity, leading to a decrease in the blood flow (von Ardenne 1980; Jain et al. 1984). These additional mechanisms unquestionably contribute to the further accumulation of lactic acid in the tumor and the reduction in the tumor tissue pH.

The changes in the tumor tissue pH following glucose administration have been studied by various techniques, including microelectrode and, more recently, NMR (nuclear magnetic resonance) techniques. The magnitude and the speed of the pH reduction depends on the amount of glucose administered. In typical experiments in which a glucose dose of 5 or 6 g/kg is given intraperitoneally to the tumor-bearing animals, the tumor tissue pH decreases slowly, reaches a minimum value in 60 to 90 minutes, and very slowly returns to the original value (Voegltin *et al.* 1936; Rhee *et al.* unpublished data). The recovery appeared to be quicker in small tumors. Minimum pH values obtained in various animal tumors are listed in Table 2, which indicates an average drop of approximately 0.5 pH units after glucose administration.

Table 2.
The effect of hyperglycemia on the rodent tumor pH.
Glucose was given intraperitoneally.

Tumor	Before	Glucose (mg/g)	After	Investigators
Rat hepatoma	6.99 (6.81–7.10)	6.0	6.42 (6.40–6.45)	Kahler and Robertson (1943)
Rat hepatoma	7.02 (6.72–7.22)	6.0	6.73 (6.35–6.99)	Kahler and Robertson (1943)
Mouse Sarcoma	7.0	5.0	6.6	Naeslund and Swenson (1953)
Rat hepatoma	6.96 ± 0.17	6.0	6.46 ± 0.22	Eden *et al.* (1955)
Rat sarcoma	6.95 ± 0.25	6.0	6.55 ± 0.27	Eden *et al.* (1955)
Rat lymphosarcoma	7.00 ± 0.20	6.0	6.50 ± 0.30	Eden *et al.* (1955)
Rat sarcoma	7.04 ± 0.11	6.0	6.67 ± 0.14	Eden *et al.* (1955)
Rat Harderian-gl.ca	7.00 ± 0.11	6.0	6.54 ± 0.23	Eden *et al.* (1955)
Rat sarcoma	7.01 ± 0.16	6.0	6.63 ± 0.22	Eden *et al.* (1955)
Rat fibrosarcoma	6.83 ± 0.24	6.0	6.48 ± 0.28	Eden *et al.* (1955)
Rat hepatoma	7.06 ± 0.22	6.0	6.62 ± 0.27	Eden *et al.* (1955)
Rat TVIA 1.0–2.5g	7.0 (6.8–7.1)*	Con.Inf.	6.5 (6.0–7.0)	Jahde and Rajewsky (1982)
Rat TVIA 4.0–6.0g	6.9 (6.7–7.1)*	Con.Inf.	6.1 (5.5–6.7)	Jahde and Rajewsky (1982)
Rat Walker 256 ca	6.98 ± 0.13	6.0	6.0	Jain *et al.* (1984)

* Continuous infusion

It has been predicted that the reduced tumor tissue pH enhances the tumor response to hyperthermia (von Ardenne 1969; von Ardenne *et al.* 1980). Urano *et al.* (1983) investigated the tumor response to 45.5 °C hyperthermia with or without glucose administration. The FSa-II tumors in animal feet were treated when tumors reached an average diameter of 4 or 8 mm and the TG time was studied. A glucose dose of 5 g/kg or 10 g/kg was given 1 hour before hyperthermia. Exponential dose response curves with small shoulders were found for tumors receiving no glucose. The dose response curves of tumors pretreated with glucose were steeper than those of tumors receiving no pretreatment, indicating that the preadministration of glucose enhanced the thermal response of the FSa-II tumor (Fig. 8). The enhancement depends on the glucose dose and the tumor size. The glucose enhancement ratio (ER) is calculated as a ratio of the treatment time to result in a TG time of 20 days for control tumors to that for glucose-treated tumors (Table 3). The ER is greater for the 8 mm tumor than for the 4 mm tumor. This is apparently attributed to a greater pH effect at the lower pH level. For an example, the ER obtained by reducing 0.5 pH unit from 7.4 to 6.9 is less significant than the ER obtained by reducing the same pH unit from 6.9 to 6.4 (see Chapter 4). Simultaneously, the RD_{50} (the treatment time to induce a loss of one toe or greater reaction in one-half of the treated animals) of the murine foot was studied with or without glucose administration. Glucose also enhances the foot reaction, but less extensively than the tumor response with a resultant therapeutic gain (Table 3).

   b. Carbon dioxide and other methods.   Inhalation of air containing excess CO_2 reduces blood and tissue pH and, as a result, may enhance thermal response of various tissues. The enhanced thermal response following the forced breathing of mixed gas containing 15 per cent CO_2 and 85 per cent air has been reported (Urano *et al.* 1980). This level of excess CO_2 in breathing air reduced blood pH value to 6.83 ± 0.19 from 7.31 ± 0.04 in normal animals. Animals with 800 mm³ FSa-II tumors were forced to breathe this mixed gas for 15 minutes before and during hyperthermia at 43.5 °C. Control animals received no specific treatment. Cell survival curves determined by lung colony assay demonstrated D_0 values of 6.3 and 5.7 for control tumors and tumors treated with excess CO_2, respectively (Fig. 9). The ER is 1.10.

   Comparable experiments were performed using immunogenic FSa-I tumors which allows for the determination of TCD_{50}. Simultaneously, the effect of excess CO_2 on the foot reaction was studied (Urano and Gerweck, unpublished

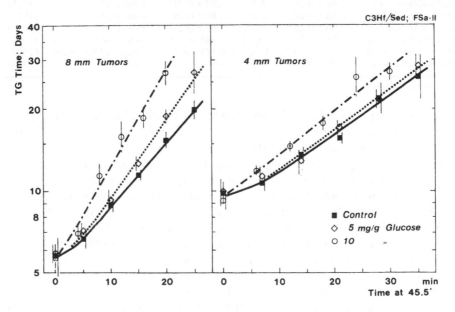

Figure 8A: Effect of hyperglycemia on the response to 45.5 °C of the Fsa-II tumor. Animals with 4 or 8 mm tumors were injected 0, 5.0 or 10.0 g/kg of glucose intraperitoneally and received local hyperthermia one hour later. Right and left panels indicate 8 and 4 mm tumors, respectively (Urano *et al.* 1983 with permission).

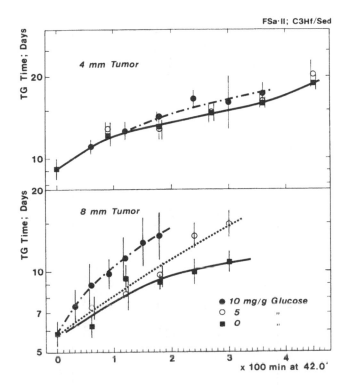

Figure 8B: Effect of hyperclycemia on the response to 42.0 °C of the FSa-II tumor. Treatments were given as mentioned in Figure 8A. Upper and lower panels indicate 4 and 8 mm tumors, respectively. Glucose enhances thermal response, particularly the large tumor at a lower temperature (Urano *et al.* 1983 with permission).

data). TCD_{50} values of the 6 mm FSa-I tumor treated at 42.5 °C with or without excess CO_2 are 213 (95% confidence limit = 189–239) or 253 (219–292) min, respectively. The RD_{50} values at 42.5 °C with or without excess CO_2 are 314 (287–344) or 359 (330–390) min, respectively. The ER by the forced breathing of excess CO_2 is 1.19 for TCD_{50} and 1.14 for RD_{50}. Gullino *et al.* (1965) observed greater tissue pH decrease in the tumor (from 7.00 to 6.67) compared to the subcutaneous tissue (from 7.38 to 7.13) in animals breathing 10 per cent CO_2 in air for 5 hours. However, the aforementioned experiments failed to demonstrate a differential response between normal and tumor tissues. The cause of failure is most likely attributed to non-uniform pH distribution in the tumor tissue.

Gullino *et al.* (1965) studied two other potential methods to reduce tumor tissue pH, i.e. forced feedings of ammonium chloride and administration of sodium bicarbonate in drinking water. Four forced feedings of ammonium chloride (75 mM/kg in 0.5 ml water each dose) given to a rat with Walker carcinoma 256 at one-hour intervals equally reduced the pH in the interstitial fluid of the tumor and in the interstitial fluid of the subcutaneous tissue.

Table 3

Effect of glucose administration on the thermal response of 2 differently sized FSa–II tumors and the foot and on the therapeutic gain.

Glucose dose (mg/g)	Tumor				Foot		Therapeutic gain for	
	4 mm		8 mm					
	T_{20}* (min)	ER	T_{20} (min)	ER	RD_{50} (min)	ER	4 mm	8 mm
at 45.5 °								
0	26.7 ± 3.6†	1.0	25.1 ± 2.8	1.0	36.3 (34.0–38.8)†	1.0	1.0	1.0
5	25.3 ± 3.2	1.06	21.1 ± 2.7	1.19	31.4 (29.5–33.4)	1.16	0.91 ± 0.13	1.03 ± 0.15
10	20.1 ± 3.2	1.33	16.1 ± 1.8	1.56	29.4 (26.5–32.5)	1.23	1.08 ± 0.20	1.27 ± 0.19
			T_{10}‡ (min)	ER				
at 42.0 °								
0			225	1.0	680 (630–720)	1.0		1.0
5			165	1.36	520 (470–560)	1.32		1.03
10			90	2.5	480 (410–560)	1.42		1.76

* T_{20} or T_{10}, treatment time at elevated temperature to prolong TG time to 20 or 10 days, respectively.

† Mean ± 95% confidence limits.

‡ Dose-response curves fitted by eyes.

The acidifying action of ammonium chloride is attributed to the metabolism of 2 NH_4^+ ions to form 1 mole of urea, 1 mole of water, and 2 H^+. The administration of sodium bicarbonate (0.1 per cent) in the drinking water reduced the pH in the tumor interstitial fluid (from 7.11 to 6.84) without affecting the pH in the interstitial fluid of the subcutaneous tissue. This is observed in tumor-bearing rats which have drunk water containing sodium bicarbonate for 10 days. Although the administration of ammonium chloride or sodium bicarbonate has been proven to reduce tumor tissue pH, these methods have not been investigated as a method to enhance the tumor response to hyperthermia.

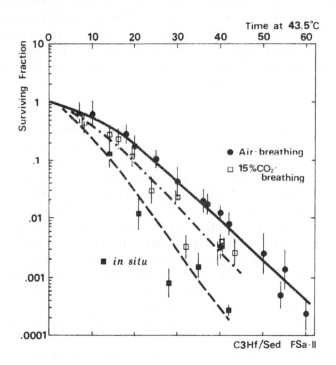

Figure 9: Cell survival curves of the FSa-II tumor cells treated at 43.5 °C *in vivo*. The 800 mm³ tumors in the feet were treated 15 minutes after the initiation of forced breathing of the mixed gas containing 15 per cent CO_2 and 85 per cent air, or 15 minute after being sacrificed. D_0 values for heat alone, heat with excess CO_2, and heat after being sacrificed are 6.3, 5.7, and 4.4 minutes, respectively (Urano *et al.* 1980 with permission).

c. Vascular Occlusion into the tumor. Vascular vessels can be occluded to cease blood circulation into the tumor. Extensive anoxia in the tumor tissue is an immediate result. Glucose may be rapidly metabolized to lactic acid which may be accumulated in the tumor without being removed. Nutrients may be consumed rapidly. Recent NMR studies confirmed rapid consumption of ATP and a decrease in the tumor tissue pH after vascular occlusion

(Gerweck et al. personal communication).

It has been demonstrated that vascular occlusion in the tumor extensively enhances the tumor response to hyperthermia. Figure 9 also demonstrates the enhanced effect on hyperthermia of vascular occlusion, together with the effect of respiration of excess CO_2 (Urano et al. 1980). Animals with 800 mm^3 FSa-II tumors were treated at 43.5 °C 15 minutes after being sacrificed and cell survival curves were obtained. The D_0 of the survival curve is reduced to 4.4 minutes from 6.3 minutes. The ER is 1.4 which is greater than the ER of 1.10 obtained by the forced breathing of excess CO_2. Hill and Denekamp (1978) observed local control of tumors treated at 44.8 °C for 15 minutes with blood vessels occluded for 80 minutes. This experiment indicates that tumor blood vessels must be occluded during hyperthermia to obtain thermal enhancement. No tumor control is observed when the tumor blood vessels were occluded before or after hyperthermia. The enhancement of tumor response by vascular occlusion has been demonstrated by other investigators (Crile et al. 1963; Suit et al. 1977).

2. Chemical modifiers

Some chemical compounds can enhance thermal response in vitro, although very few studies have been performed in animals. One of the heat sensitizers is 5-thio-D-glucose (5–TDG), An antimetabolite of D-glucose. It gives lethal damage to hypoxic cells (Song et al. 1976) and enhances thermal response of these cells (Kim et al. 1978; Song et al. 1979). The 5-TDG inhibits the anaerobic glycolytic pathway, thus deprives energy needed to sustain hypoxic cells. Furthermore, this agent, if given together with hyperthermia, may reduce the oxygen uptake by hypoxic cells, resulting in a higher degree of hypoxia. Mammalian cells are sensitive to heat in energy-deprived conditions.

Reinhold and Van den Berg-Blok (1981) studied the effect of 42.0 °C hyperthermia given alone or in combination of 5-TDG on the microcirculation in a sandwich tumor model (a tumor growing in a thin chamber implanted subcutaneously in the back of rats). The chamber walls are made of transparent mica, thus the tumor tissue is visible by eye, or under a microscope as a thin layer. The tumor was rat rhabdomyosarcoma BA 1112. The effect of 42.0 °C hyperthermia on the microcirculation was substantially enhanced by 5-TDG. Further studies are required to further characterize and optimize the effect of this agent on the response of tumor and normal tissues to elevated temperatures. Another antimetabolite, 2-deoxy-glucose might be a potential heat sensitizer as well.

Kim et al. (1978) noticed that the 5-TDG is a potent antifertility agent and examined the heat sensitizing effect of other potent antifertility agents including lonidamine (dichlorophenyl-methyl-lH-indazole-carboxylic acid) and gossypol (Kim et al. 1984, 1985). The latter blocks glycolysis and, as a result, deprives the energy from the cells. Lonidamine interferes with lactate transport with a resultant accumulation of lactic acid in the cells. Another agent, quercetin, also sensitizes mammalian cells to heat by inhibiting lactate transport (Kim et al. 1984). Among these agents, lonidamine was investigated in an animal tumor system. Kim *et al.* (1984) reported that 84 and 45 per cent of Meth-A fibrosarcomas in Balb/c mice were controlled by lonidamine (50 mg/kg) plus hyperthermia (41.6 °C for 90 min) and by hyperthermia alone, respectively.

Other types of heat sensitizers are the local anesthetics, such as lidocaine or procaine. The sensitizing effect is well-documented in cultured cells and the interaction is assumed to occur at cell membrane level (Yau and Kim 1980). Thermal enhancement effect of these agents has been shown in an animal tumor system. Yatvin *et al.* (1979) observed the enhanced effect at 43.5 °C, but not at 42.0 °C . Unlike local anesthetics, nembutal (Pentobarbital), a general anesthetic commonly used in animal studies, does not modify the thermal effect. On the other hand, a general anesthetic, urethane (ethylcarbamate), which has a weak chemotherapeutic effect against leukemia, shows enhanced cytotoxicity at elevated temperatures (Urano *et al.* 1981). This enhancement may be due to thermal enhancement of the cytotoxic effects of the agent rather than heat sensitization by the agent.

It is also well-documented that polyamines enhance the thermal effect on mammalian cells (Ben-Hur *et al.* 1978; Gerner *et al.* 1980). The mechanism of this interaction is suggested to be due to the polyamine action on the cell membrane. Again, the effectiveness of these chemicals on tumor or normal tissue response must await further studies. Methylglyoxal Bis (quanylhydrazone) which inhibits at least two enzymes involved in polyamine metabolism is another heat sensitizer. Its effect on cultured cells and on the murine skin has been demonstrated (Leith 1982; Miyakoshi and Heki 1983). It is of interest that this agent can inhibit the development of thermotolerance in the murine skin. Some agents can protect thermal cell killing *in vitro* (Brenner *et al.* 1981; Fisher *et al.* 1982; Henle *et al.* 1983). However, no studies have been performed in experimental animals.

E. STEP-DOWN HEATING AND THERMOTOLERANCE.

Treatment temperature itself modifies the thermal response of cells and tissues. This effect may be sensitization or protection depending on the temperature applied and the timing of the second heating. The effect of hyperthermia given at temperatures above 43.0 °C can be enhanced by hyperthermia given at temperatures below 43.0 °C, if cells or tissues exposed to heat above 43.0 °C are immediately transferred to the temperature below 43.0 °C. This has been called step-down heating. On the other hand, the cells or tissues become resistant to an elevated temperature as a result of a prior or continuous exposure to hyperthermia. This phenomenon occurs at any elevated temperature and is called thermotolerance. Extensive discussion on the latter phenomenon will be reserved for another volume of this series, since this is critical for clinical hyperthermia.

1. Step-Down Heating (SDH)

It has been demonstrated in cultured mammalian cells that the effect of an elevated temperature above 43.0 °C is enhanced by the subsequent hyperthermia below 43.0 °C when the second heat treatment is applied immediately or shortly after the first hyperthermia (Henle *et al.* 1978; Joshi and Jung, 1979; Henle 1980). This SDH effect has been observed in animal normal and tumor tissues as well.

The substantial prolongation of the TG time is observed when the 4 mm FSa-II tumor in the murine foot is first treated for 15 minutes at 45.5 °C followed by a second hyperthermia treatment for 2 hours at 41.0 °C which is given immediately thereafter (Urano and Kahn 1983, Fig. 10). The TG time after the first hyperthermia alone was 11 days and the SDH prolonged it to 21.5 days. The effect of the SDH decreases with increasing treatment interval between the first (higher temperature) and the second (lower temperature) treatments, and is lost within 16 hours. This decay of the SDH effect is very similar to that observed in cultured cells, except that a 2-hour delay of the application of SDH did not affect the SDH sensitization in the tumor (also in the normal tissue), although the same short delay decreased the magnitude of sensitization *in vitro*. This difference may be due to physiological changes in the heated tissue such as decreased blood flow or tissue pH. Dose response curves for the FSa-II tumors treated at 45.5 °C with or without SDH are shown in Fig. 11, together with the effect of SDH on thermotolerance

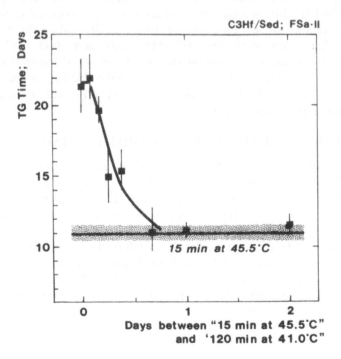

Figure 10: Effect of treatment interval between initial high temperature (for 15 min at 45.5 °C) and second low temperature (for 2 hours at 42.0 °C) treatments. The SDH enhanced the initial high temperature treatment, but the enhancement decreased with increasing time interval (Urano and Kahn 1983 with permission).

(which will be discussed in the next section). The ER due to SDH of 120 minutes at 41.0 °C is calculated as a ratio of treatment time to induce a TG time of 30 days; it is 1.71 (24/14 min). Urano and Kahn (1983) investigated the SDH effect on the murine foot response and confirmed that the effect was identical in both tumor and normal tissues. This indicates that the SDH results in no therapeutic gain.

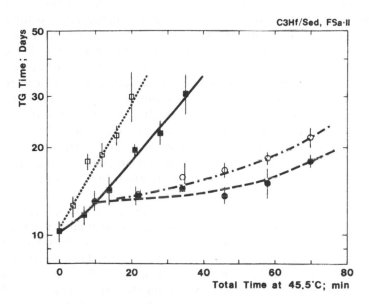

Figure 11: Effect of the SDH on the response of the FSa-II tumors to 45.5 °C. Tumors were treated for various times at 45.5 °C and immediately transferred to 42.0 °C. The SDH was for 2 hours. Solid square are tumors treated at 45.5 °C; open square indicate tumors treated at 45.5 °C and with SDH. The ER by the SDH at the TG time of 20 days is 1.7. Circles indicate that tumors treated for 10 minutes at 45.5 °C received the second treatment for various times at 45.5 °C 24 hours after the first treatment. Tumors treated with first hyperthermia developed substantial thermotolerance 24 hours later. Open and solid symbols are those treated with or without the SDH at the second treatment. The effect of the SDH on thermotolerant cells is very small (Urano and Kahn 1983 with permission).

The effect of the SDH depends on the temperature and duration of the SDH (second lower temperature treatment). It is demonstrated *in vitro* (Henle 1980) and in the murine foot (Urano and Kahn 1983) that a 1 °C increase in the SDH temperature halves the treatment time of the SDH. The mechanism of the SDH has been discussed by some investigators (Henle 1980; Urano and Kahn 1983). It is likely that the repair of sublethal and/or potentially lethal heat damage induced above 43.0 °C is inhibited by the SDH, resulting in a thermal enhancement.

2 Thermotolerance

Thermotolerance is a phenomenon in which cells or tissues become resistant to elevated temperatures as a result of prior or continuous exposure to hyperthermia. In the early sixties, Crile (1963) already observed this phenomenon in animal normal and tumor tissues. Exposure of the murine foot for 30 minutes at 44 °C protected against a subsequent treatment of 90 minutes at 44 °C given 24 hours later. Similar protection was also observed in S91 DBA mouse melanoma. Recent extensive research disclosed that cells which acquired thermotolerance exhibited a decreased slope of the cell survival curve with or without a change in the shoulder, compared with the survival curve of non-preheated cells (Henle and Dethlefsen 1978). At temperatures below 43 °C (below the breaking point on the Arrhenius plot), thermotolerance develops during a continuous heat treatment and forms a resistant tail in the cell survival curve. Similar tails are observed in the dose response curves between the TG time and the treatment time (see Fig. 1). Above 43.0 °C, thermotolerance usually develops after the first exposure, and cells or tissues become resistant to subsequent heat treatment.

Thermotolerance is a transient phenomenon. The kinetics (development and decay) of thermotolerance has been studied in cultured mammalian cells (Gerner and Schneider 1975; Henle and Dethlefsen 1978). It develops rapidly within 8 to 24 hours after an acute exposure to elevated temperatures and decays slowly over 3 to 7 days. In animal tumors, it also develops rapidly and decays slowly (Maher et al. 1981). In the FSa-II tumor it develops in 8 to 16 hours after a 7.5 minute treatment at 45.5 °C and decays slowly over the next 7 days (Urano and Kahn 1986). The magnitude of thermotolerance appears to increase with increasing first heat dose. The time to reach the maximum thermotolerance also depends on the initial heat dose. Nielsen and Overgaard (1982) demonstrated these relationships following the various first dose at 43.5 °C (Fig. 12). A fixed second dose was given at various times thereafter. They observed maximum tolerance at 2, 8 and 26 hours after the first treatment of 2.5, 15 and 45 minutes at 43.5 °C, respectively.

It may be of interest to compare the kinetics and the magnitude of thermotolerance in the normal and tumor tissues. It has been shown in some rodent normal tissues including ear, foot, tail and whole-body response, that the decay of thermotolerance is incomplete over a treatment interval of $\geq$ 7 days, while thermotolerance decays completely in murine tumors (Urano 1986).

The thermotolerance ratio (TTR), or the magnitude of thermotolerance can be expressed in various ways e.g., as a ratio of D_0 values of the survival curves for preheated and non-preheated cells. For studies using a single endpoint, such as RD_{50}, it can be expressed as a ratio of second doses to induce a specified reaction (see Chapter 5 for detailed discussion of the TTR). Namely,

$$TTR = \frac{D_2}{D_s - D_1}$$

where D_s is the single dose to induce the specified reaction, and D_1 and D_2 are the first and second doses, respectively. In Fig. 13, TTRs determined at 24 hours after the first dose (i.e. maximum or near maximum thermotolerance) are shown as a function of initial dose (Urano 1986). It is notable that the magnitude of thermotolerance in three different murine tumors is greater than that in the murine foot. Regarding the normal tissues, the magnitude of thermotolerance in the murine ear (Law *et al*. 1979) is very similar to that in the murine foot (Rice *et al*. 1982).

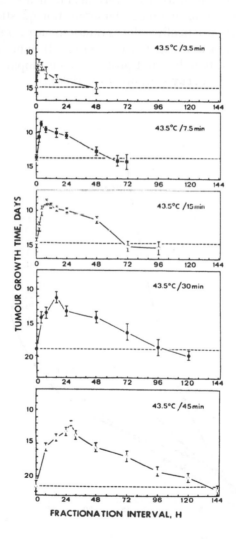

Figure 12: The effect of various first treatment times on the magnitude of thermotolerance and the time to reach maximum tolerance. Mouse mammary carcinoma in the foot received various initial treatments at 43.5 °C and then the second treatment for 30 minutes at 43.5 °C at various times thereafter. The magnitude of thermotolerance increases and the time to reach maximum prolongs with increasing initial treatment time (Nielsen and Overgaard 1982 with permission).

It has been shown in cultured mammalian cells that the magnitude of thermotolerance decreases with decreasing extracellular pH (Gerweck 1977, 1980; Goldin and Leeper 1981). The tumor tissue pH, as has been discussed previously, is lower compared to the normal tissue pH. In this situation, the magnitude of thermotolerance could be smaller in the tumor than in the normal tissue. However, substantial thermotolerance has been observed in tumor tissue as shown in Fig. 13. This might be the result of the non-uniform pH distribution in the tumor. Namely, cells in low pH environment are killed by the first heat treatment, and cells in normal or near-normal pH environment develop substantial thermotolerance. An interesting question is if the glucose administration which specifically reduces tumor tissue pH inhibits the development of thermotolerance. A study using the FSa-II tumor demonstrates that the glucose administration does not protect the development of thermotolerance, but facilitates its decay (Urano et al. 1984).

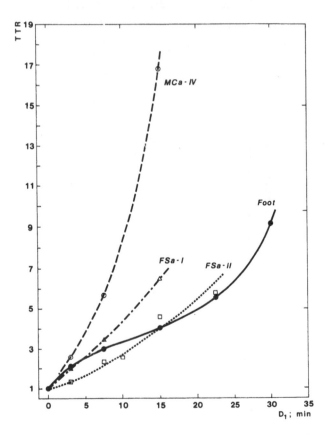

Figure 13: Thermotolerance ratio (TTR) as a function of initial treatment time at 45.5 °C for various murine tumors and the murine foot. The second treatment was given 24 hours after the first treatment and the TTR was obtained at a given endpoint (the treatment time to induce a given TG time). See the equation shown in the text for the calculation of TTR (Urano 1986 with permission).

Recent investigations have disclosed that the development of thermotolerance is associated with an increased synthesis of some proteins called heat shock proteins (see Chapter 3). This increased protein synthesis is found in a murine tumor treated at elevated temperatures (Li and Mak 1985). Intensive heating of a tumor damages tumor vasculature with a resultant vascular occlusion and pH reduction. These pathophysiological changes in the tumor temporarily increases thermal sensitivity of a rodent tumor, although thermotolerance developed subsequently and overcame the increased sensitivity within several hours (Rhee *et al.* 1982).

F. EFFECT ON IMMUNE REACTION.

From the beginning of hyperthermia research, many investigators have focused on the relationship between thermal effects and the host immune reactions. These studies can be classified into three categories; the effect of whole-body hyperthermia on the host immune reaction, the effect of heat-killed tumor cells on the proliferative tumor cells, and the effect of non-specific immune stimulant at elevated temperatures on the tumor tissue. Although some investigators observed an abscopal effect in patients with multiple metastases (treatments of a tumor leads to a disappearance of an untreated tumor) following hyperthermia (Goldenberg and Langner 1971), it must be noted that only a few types of human tumors are strongly immunogenic. In animal tumor systems, most spontaneous tumors are weakly or non-immunogenic, while most chemically-induced tumors are moderately immunogenic.

1. The effect on the host immune reaction
MacDonald (1977) studied the effect of hyperthermia on the functional activity of murine cytotoxic T cells. The T cells were generated *in vitro* in mixed leukocyte cultures with the use of an equal number of C57BL/6 spleen cells and irradiated DBA/2 spleen cells. The mixed leukocyte cultured cells were exposed to temperatures above 42.0 °C for various times, and cytotoxicity on ^{51}Cr-labeled mastocytoma cells was examined. The degree of inactivation depended on the temperature and the treatment time. The treatment time to reduce the lytic activity of cytotoxic T cells to a 10 per cent level was 21 minutes and 9 minutes at 43 °C and 44 °C, respectively (Fig. 14). This time–temperature relationship suggests the same activation energy for cytotoxic T cells as that for other mammalian cells and tissues; namely, a 1 °C increase in the treatment temperature halves the treatment time to induce a specific biological effect. Split dose experiments using two 10-minute treatments at 43 °C separated by 3 hours at 37 °C suggest that thermotolerance also develops in the cytotoxic T cells. Harris (1976) also reported an inhibitory effect of 43 °C hyperthermia on the cytotoxic T cells.

A local curative heating of rabbit VX-2 tumors protects against the subsequent transplantation of the same tumor cells (Dickson and Shah 1983). These animals also show an increase in the skin response to challenge with tumor extract. However, this increased host-immune response was abrogated when whole-body hyperthermia followed local tumor heating. This study supports the inhibitory effect of hyperthermia on cell-mediated and humoral immune responses.

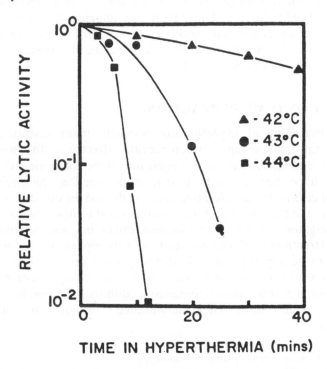

Figure 14: Relative lytic activity of cytotoxic T cells as a function of treatment time at elevated temperatures (MacDonald 1977 with permission).

Tumor cells injected intravenously in mice may lodge in the lung and form macroscopic colonies within 2 to 3 weeks. This lung colony forming ability may be increased when the host immune reaction is suppressed. Three daily whole-body hyperthermia of 60 minutes at 41.5 °C (rectal temperature) each increased the number of lung colonies of the immunogenic FSa-I tumor cells, approximately 3 fold, compared to the number of colonies formed in the non-treated animals (Fig. 15), indicating that whole-body hyperthermia suppressed the host-immune response (Urano *et al*. 1983). However, this kind of reaction was not observed against the non-immunogenic FSa-II tumor.

2. Heat-killed tumor cells
It has been demonstrated that lethally *irradiated* tumor cells transplanted intraperitoneally or subcutaneously can immunize the recipient against the subsequent challenge of the same tumor cells. Mondovi *et al*. (1972) demonstrated that similar immunization was obtained by lethally *heated* Ehrlich ascites tumor cells, which were given by two intraperitoneal injections sepa-

rated for 20 days. The same untreated tumor cells were transplanted intra-peritoneally 26 days after the second immunization. The tumor take rate was significantly less in the animals immunized with *heat*-killed cells than in animals receiving no immunization. Animals treated with *radiation*-killed tumor cells were less significantly immunized than animals treated with *heat*-killed tumor cells.

Effect of WBH on Lung Colony Forming Ability

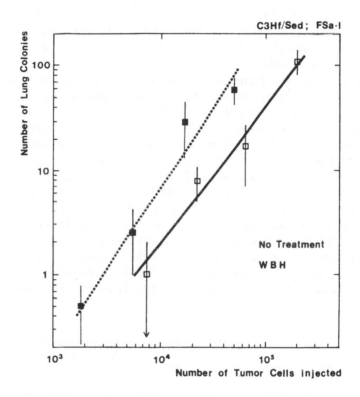

Figure 15: Relationship between the number of lung colonies and the number of FSa-I tumor cells injected. Open or solid symbols indicate that recipient animals received no pre-treatment or whole-body hyperthermia 24 hours prior to injection, respectively (Urano *et al.* 1983).

A quantitative study for immunogenicity of FSa-I tumor cells killed by heat or irradiation was performed by Suit *et al.* (1977). Animals were immunized with three weekly injections of heat- or radiation-killed tumor cells. These cells were obtained by an *in vitro* irradiation of 100 Gy, or an *in vitro* heating for 240 minutes at 43.5 °C. The TD_{50} values in non-treated animals, animals immunized with *heat*-killed cells, and animals immunized with *radiation*-killed cells were 9×10^3, 5.1×10^4 and 4.4×10^6 cells, respectively. These results indicate that immunization by *radiation*-killed cells was greater than that by *heat*-killed cells. These investigators also examined the effects of another immunization procedure. Animals received a transplantation of FSa-I cells into the leg. Presumably, animals have been immunized with the establishment of the tumor. When the tumors reached an average diameter of 6 to 7mm, they were treated with heat (120–135 min at 43.5 °C), irradiation (50 or 60 Gy), or amputation. Each treatment controls the tumor with $\geq$ 95 per cent probability. The TD_{50} assays performed 1 to 3 weeks after treatment showed similar results as those in the aforementioned experiments.

This second immunization procedure was also used by Dickson and Shah (1983). Curative local heating of established VX-2 rabbit tumors increased the skin response to challenge with tumor extract. An increase in the antitumor antibody titers was also demonstrated. No comparison was made with radiation-killed tumor cells.

A comparison in the degree of immunization and the effect of hyperthermia on the host immune response was made between immunogenic FSa-I tumors, and non-immunogenic FSa-II tumors (Urano et al. 1983). Tumors were transplanted into the animal leg. When tumors reached an average diameter of 6 or 10mm, tumor-bearing legs were amputated. Immediately after amputation, animals were subjected to a single whole-body irradiation of 6.0 Gy, or 3 daily whole-body hyperthermia for 60 minutes at 41.5 °C each, and then received an i.v. injection of a given number of tumor cells. The degree of immunization was evaluated by comparing the number of lung colonies formed (Table 4). The number of lung colonies was smaller in animals which had the FSa-I tumor than in animals which had no tumor, and decreased with increasing tumor size. This indicates that immunity against the FSa-I tumor developed with tumor growth. Unlike the FSa-I tumor, no immunity developed against the non-immunogenic FSa-II tumor. The data from this

Table 4
Effect of whole-body irradiation (WBI) and whole-body hyperthermia (WBH) on the lung colony formation in animals which had tumors (tumor bearing) and animals which received no previous transplantation (non-tumor bearing). An intravenous injection of 1.5×10^4 tumor cells was given 1 day after the last treatment.

Treatment	Number of lung colonies (mean ± S.D.)	
	FSa-I*	FSa-II*
Non-Tumor Bearing		
No treatment	31 ± 14	32 ± 16
WBI	85 ± 25	56 ± 34
WBH	55 ± 18	27 ± 16
Tumor Bearing (6mm)		
Amputation alone	11 ± 5	22 ± 3
Amputation + WBI	51 ± 36	41 ± 5
Amputation + WBH	18 ± 12	15 ± 5
Tumor Bearing (10mm)		
Amputation alone	2 ± 2	24 ± 5
Amputation + WBI	4 ± 6	34 ± 19
Amputation + WBH	3 ± 3	20 ± 11

* The FSa-I is chemically induced and immunogenic, while the FSa-II is a spontaneous fibrosarcoma and non-immunogenic.

experiment further indicate that whole-body hyperthermia does not significantly suppress the developed immunity, although whole-body irradiation does. These results, together with others, suggest that the antigenicity of the tumor is a critical factor in the establishment of the host immune reaction and in the thermal effect on the immune reaction.

3. Immune stimulants and hyperthermia

Formalin- or heat-killed *Corynebacterium parvum (C. parvum)* stimulates reticuloendothelial system and augments macrophage function (Adam and Scott 1973; Woodruff *et al*. 1974). It retards growth of some animal tumors and occasionally causes complete regression of established tumors (Milas 1974). It can be classified as a non-specific immune stimulant.

It has been demonstrated that *C. parvum* enhances thermal response of tumor and normal tissues. This enhancement is observed only *in vivo*, indicating that the enhancement of *C. parvum* may be mediated by host-tumor relationship. The TCD_{50} value at 43.5 °C of the immunogenic FSa-I tumor was 92.6 minutes and was reduced to 25.0 minutes by *C. parvum* administered 3 days before hyperthermia. The enhancement ratio (ER) expressed as a ratio of TCD_{50} (control) to TCD_{50} *(C. parvum)* was 3.70. The RD_{50} values at 43.5 °C of the murine foot with or without *C. parvum* administration were 92 or 133 minutes, respectively. The ER was 1.45. Although *C. parvum* enhanced thermal response of both normal and tumor tissues, the therapeutic gain factor for the FSa-I tumor was 2.55 (3.70/1.45) (Urano *et al*. 1978). The enhancement was not as dramatic against the FSa-II tumor compared to the FSa-I, but the therapeutic gain was still observed. The treatment time to induce TG time of 20 days with or without *C. parvum* administration was 55 or 103 minutes, respectively. The ER was 1.89, which is not only due to the enhancement, but also due to the additive effect, i.e. *C. parvum* itself prolonged the TG time. The therapeutic gain factor for the non-immunogenic FSa-II tumor was 1.29 (Urano *et al*. 1979). The ER for non-immunogenic C3H mouse mammary carcinoma was ≥ 2.0, resulting in a therapeutic gain factor of ≥ 1.38 (Urano, unpublished data, Fig. 16). A suggested mechanism of thermal enhancement by *C. parvum* is that *C. parvum*-induced macrophages attack heated tissues more severely than normal macrophages (Urano *et al*. 1984).

Currently, the anti-tumor effects of interferon and interleukin-2 have been extensively studied. It has been shown that antiviral activity and growth inhibitory effect of interferon are enhanced at moderately elevated temperatures (39.0 °C or 40.0 °C) (Heron and Berg 1978; Delbruck *et al*. 1980). Yerushalmi *et al*. (1982) demonstrated that the mean survival time of C57BL mice bearing Lewis lung carcinoma was significantly prolonged by combined interferon and local heat treatments. It is notable that the cytotoxicity of interferon has been observed in cultured mammalian cells where no host-mediated reaction is involved (Fleischmann *et al*. 1986). This definitely suggests a potential for interferon to be used as an anti-tumor agent in combination with hyperthermia. Simultaneously, it raises a question that combined treatments may cause or enhance the side-effect of the agent.

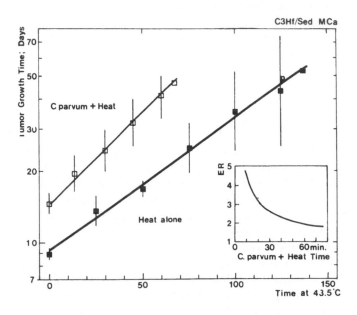

Figure 16: Effect of *C. parvum* on the response of a mouse mammary carcinoma to 43.5 °C hyperthermia. The TG time is plotted as a function of treatment time. *C.parvum* showed both additive effect and enhancement. Inserted figure shows the ER as a function of treatment time at 43.5 °C. The treatment time is for combined *C. parvum* and hyperthermia (Urano *et al.*, unpublished data).

G.   EFFECT ON METASTASIS

A critical question for a new treatment modality is whether the treatment may induce lymphogenic and/or hematogenic metastasis. This question has been asked for local and whole-body hyperthermia. Increased blood flow has been demonstrated in the normal tissue at an elevated temperature, although blood flow in the tumor increases at relatively low temperature hyperthermia and does not increase at 43.0 °C or above (Bicher *et al.* 1980; Eddy 1980). Vascular damage in the tumor is commonly induced at elevated temperatures, frequently leading to the obstruction of blood vessels with resultant increase in the hypoxic cell fraction (Song *et al.* 1982; Urano and Kahn 1983). These vascular changes may influence the incidence of tumor metastases. The metastasis incidence at an elevated temperature might be influenced by the membrane damage induced by hyperthermia. It is demonstrated that host-immune status plays a major role in the metastasis frequency. Although experimental results obtained are still controversial, these results suggest that local hyperthermia (LH) does not enhance metastasis frequency, but whole-body hyperthermia (WBH) may increase metastasis frequency. Table 5 summarizes the data reported on the metastasis incidence following hyperthermia alone, including LH and WBH.

Table 5
Metastasis frequency following hyperthermia alone, local heat (LH), or whole-body hyperthermia (WBH)

Tumor	LH	WBH	Investigator	
Allogeneic				
VX-2 carcinoma (rabbit)	N.C.*	Increase	Dickson and Muckles	(1972)
Yoshida sarcoma (rat)	Increase	–	Dickson and Ellis	(1974)
Syngeneic				
Immunogenic				
Lewis lung carcinoma (mouse)	N.C. or Increase	Increase	Yerushalmi	(1976)
FSa-I (mouse)	N.C.	N.C.	Urano *et al.*	(1984)
Lewis lung carcinoma (mouse)	–	N.C.	Oda *et al.*	(1985)
Non-Immunogenic				
C3H carcinoma (mouse)	Increase	–	Walker *et al.*	(1978)
FSa-II (mouse)	N.C.	Increase	Urano *et al.*	(1984)
CA SQD (mouse)	N.C.	–	Hill and Denekamp	(1982)
Immunogenicity Unknown				
Dunn osteogenic sarcoma (mouse)	N.C.	–	Hahn *et al.*	(1979)
KHT mammary carcinoma (mouse)	N.C.	–	Marmor *et al.*	(1979)
Hepatoma–134 (mouse)		Increase	Oda *et al.*	(1985)

* No change

    The LH alone does not enhance the metastasis incidence except against Yoshida sarcoma and a C3H mouse carcinoma. Dickson and Ellis (1974) reported that 32 per cent of the animals bearing Yoshida sarcoma showed lymph-node metastasis. This incidence increased to over 50 per cent following LH at 42.0 °C for 24 hours or longer. Metastases in unusual locations were also noted. Walker *et al.* (1978) reported that the incidence of metastasis following curative LH was 27 per cent, while it is 11 per cent following curative local irradiation. All other studies failed to demonstrate significant promotion of metastasis by LH. Urano *et al.* (1983) noticed a slight increase in the metastasis incidence following LH, which was not statistically significant.

    In order to treat animal tumors in the water bath, each animal must be restrained in a holder. This is often done without anesthesia. A question is whether such restraint may cause an increase in the metastases incidence. Ando *et al.* (1987) assigned animals with 8 mm FSa-II tumors in their feet into three groups; animals receiving no treatment, animals receiving amputation alone, and animals kept in individual holders without receiving hyperthermia. No anesthesia was given at the time of treatment. Animals in the latter group also received amputation 24 hours after hyperthermia. The metastasis incidence in animals receiving no treatment, amputation alone, and restrained in holders without hyperthermia was 30.3 (10/33), 29.3 (12/41), and 67.4 (29/43) per cent, respectively. In comparable experiments, animals received various lengths of heat treatment at 41.5 °C (from 60 to 240 min), or 45.5 °C (from 5 to 20 min.) They showed no increase in the metastasis incidence. The incidence was between 20 per cent and 40 per cent. These results indicate that the animal restraint causes an increase in the metastasis incidence and hyperthermia did not increase this incidence.

It rather decreased it from 67 per cent to a maximum of 40 per cent. They also show that glucose administered 60 minutes before hyperthermia did not affect the metastasis incidence.

Host–tumor relationships influence the incidence of metastasis. Animals transplanted with the immunogenic FSa-I tumor obtain immunogenicity against the retransplantation of the same tumor cells, and whole-body hyperthermia does not suppress this established immunity (Table 4). This explains why whole-body hyperthermia does not increase the incidence of metastasis of the immunogenic FSa-I tumor. Increased incidence of metastasis of Lewis lung carcinoma may be attributed to the rapid development of metastasis and early application of whole-body hyperthermia. Namely, whole-body hyperthermia was applied before the establishment of anti-tumor immunity. This assumption might be supported by the evidence that the increase in the metastasis incidence was observed when animals were heated one day after tumor transplantation, and no significant increase was seen when the animals were treated 10 days after transplantation.

Animals with non-immunogenic FSa-II tumors showed an increased incidence of metastasis following 3 daily whole-body hyperthermia of 60 minutes at 41.5 °C each. Animals were held in a wire cage without anesthesia during the time of treatment. It is not clear whether this is caused directly by whole-body hyperthermia or by restraining animals in a cage, although the animals could move about more freely in a cage compared to individual holders used for LH. Lord et al. (1981) reported that dogs treated with whole-body hyperthermia and radiation for osteosarcoma showed increased skeletal metastases. Animals were anesthetized at the time of treatment. This result indicates direct involvement of whole-body hyperthermia in the increased incidence of metastasis. It is likely that whole-body hyperthermia induces the lung damage which enhances the metastasis frequency as does cyclophosphamide or whole-body irradiation (Brown and Marsa 1978).

Baker et al. (1981) studied metastasis incidence after irradiation of KHT sarcomas and concluded that the increased incidence following irradiation was a result of the manipulations associated with irradiation. These results suggest that great caution must be exercised for handling animals in metastasis experiments and that further studies are needed before drawing any conclusions for whole-body hyperthermia.

A factor which should be discussed in this section is the heterogeneity of tumor cells (Hart and Fidler 1981). Tomasovic et al. (1982) separated clones with various metastatic potential. Clones with high metastatic potential were not inherently more or less sensitive to hyperthermia than clones with low or intermediate metastatic potential.

H.   CONCLUSION

It has been established that thermal sensitivity of mammalian cells increases with decreasing environmental pH. The measurements of the tumor tissue pH have been performed in many laboratories and confirmed that the tumor tissue pH, including rodent and human tumors, is lower compared to the normal tissue pH. Presumably, both anaerobic and aerobic glycolysis activity is high in the tumor, resulting in the accumulation of lactic acid. Average tumor tissue pH decreases with increasing tumor size, and as a result, thermal response of the tumor increases with tumor growth. It is unlikely, however, to predict uniform pH distribution with uniform thermal sensitivity.

Thermal sensitivity of the tumor depends on the treatment temperature and time. Above the breaking point on the Arrhenius plot, a 1 °C increase halves the treatment time to induce a specified biological response. Below the point, the same 1 °C increase decreases the treatment time by a factor of 3 to 4. The temperature at the breaking point appears to decrease with increasing tumor size.

Various methods have been developed to reduce tumor tissue pH with the resultant enhancement of tumor response to hyperthermia. Among them, hyperglycemia appears to be promising since it specifically reduces tumor tissue pH by accelerating glycolysis activity and reducing blood flow in the tumor. Although chemical modifiers of thermal response are briefly discussed, more extensive discussions are given elsewhere in this volume.

The effect of hyperthermia given at temperatures above 43.0 °C can be enhanced by the treatment given at temperatures below 43.0 °C, if the lower temperature treatment immediately follows the higher temperature treatment. This phenomenon has been called step-down heating (SDH). The SDH effect has been observed in both normal and tumor tissues with no resultant therapeutic gain.

The cells and tissues become resistant to elevated temperatures as a result of a prior or continuous exposure to hyperthermia. This phenomenon called thermotolerance has been observed in tumors and normal tissues. Above 43.0 °C, it develops rapidly, reaches maximum in 8 hours to 24 hours and decays slowly. The magnitude of thermotolerance and the time required to reach maximum depends on the initial heat dose. A significant difference between normal and tumor tissues in the kinetics of thermotolerance has been observed; namely, in some normal tissues, including ear, foot, tail, and whole-body response. The decay of thermotolerance is incomplete over a treatment interval $\geq$ 7 days, while it appears to be complete in the tumor. In cultured mammalian cells, the magnitude of thermotolerance decreases with decreasing extracellular pH. Although the tumor tissue pH is lower compared to the normal tissue pH, the magnitude of thermotolerance is greater in some tumors, regardless of the tumor size, than in normal tissues.

The effect of hyperthermia, particularly whole-body hyperthermia, on the host immune reaction has been studied. Whole-body hyperthermia inhibits host immune responses. Survival curves of cytotoxic T cells following hyperthermia are similar to those of other mammalian cells. It has been shown that *heat*-killed tumor cells transplanted intraperitoneally or subcutaneously can immunize the recipient against the subsequent challenge of the same tumor cells. *Corynebacterium parvum* , which stimulates reticuloendothelial system and augments macrophage function enhances thermal response of tumor and normal tissues.

Experimental results obtained for the effect of hyperthermia on metastasis are controversial. It appears to depend on the timing between transplantation and treatment(s), assay methods and the type of experimental tumors. Well-designed experiments are needed before any conclusion can be made.

REFERENCES

Adam, C. and Scott, M.I. (1973). Lymphoreticular stimulatory properties of corynebacterium parvum and related bacteria. *J. Med. Microbiol.* **6** (3), 261–274.

Ando, K., Urano, M., Kenton, L. and Kahn, J. (1987) Effect of thermochemotherapy on the development of spontaneous lung metastases. *Int. J. Hyperthermia* **3**, 453–458.

Baker, D., Elkon, D., Lin, M–L., Constable, W. and Wanebo, H. (1981). Does local X-irradiation of a tumor increase the incidence of metastases? *Cancer* **48**, 2394–2398.

Ben-Hur, E., Prager, A. and Riklis, E. (1978). Enhancement of thermal killing by polyamines I. Survival of Chinese hamster cells. *Int. J. Cancer* **22**, 602–606.

Bicher, H.I., Hetzel, F.W., Sandhu, T.S., Frinak, S., Vaupel, P., O'Hara, M. and O'Brien, T. (1980). Effects of hyperthermia on normal and tumor microenvironment. *Radiology* **137**, 523–530.

Brown, J.M. and Marsa, G.W. (1978) Effect of dose fractionation on the enhancement by radiation or cyclophosphamide of artificial pulmonary metastases. *Br. J. Cancer* **37**, 1020–1025.

Brenner, H.J., Leith, J.T., DeWyngaert, J.K., Dexter, D.L., Calabresi, P. and Glicksman A.S. (1981). Protection against hyperthermic cell killing of mouse mammary adenocarcinoma cells *in vitro* by N,N-dimethylformamide. *Radiat. Res.* **88**, 291–298.

Cavaliere, R., Ciocatto, E.C., Giovanella, B.C., Heidelberger, C., Johnson, R.O., Margottini, M., Mondovi, B., Moricca, G. and Rossi-Fanelli, A. (1967). Selective heat sensitivity of cancer cells. *Cancer* **20**, 1351–1381.

Crile, G., (1963). The effects of heat and radiation on cancers implanted on the feet of mice. *Cancer Res.* **23**, 372–380.

Delbruck, H.G., Allouche, M. and Jasmin, C. (1980). Influence of increased temperature on the inhibition of rat osteosarcoma cell multiplication «in vitro» by interferon. *Biomedicine* **33**, 239–241.

Dewey, W.C., Hopwood, L.E., Sapareto, S.A. and Gerweck, L.E. (1977). Cellular responses to combinations of hyperthermia and radiation. *Radiology* **123**, 463–473.

Dickson, J.A. and Ellis, H.A. (1974). Stimulation of tumour cell dissemination by raised temperature (42 °C) in rats with transplanted Yoshida tumours. *Nature* **248**, 354–358.

Dickson, J.A. and Shah, S.A. (1983). Immunological aspects of hyperthermia. In *Hyperthermia in Cancer Therapy*, F.K. Storm (Ed.), G.K. Hall and Co., Boston, MA, pp. 487–543.

Eddy, H.A. (1980). Alterations in tumor microvasculature during hyperthermia. *Radiology* **137**, 515–524.

Eden, M., Haines, B. and Kahler, H. (1955). The pH of rat tumors measured *in vivo*. *J. Natl. Cancer Inst.* **16**, 541–556.

Fajardo, L.F., Egbert, B., Marmor, J. and Hahn, G.M. (1980). Effects of hyperthermia in a malignant tumor. *Cancer* **45**, 613–623.

Fisher, G.A., Li, G.C. and Hahn, G.M. (1982). Modification of the thermal response by D$_2$O I. Cell survival and the temperature shift. *Radiat. Res.* **92**, 530–540.

Fleischman, W.R., Fleischman, C.M. and Gindhart, T.D. (1986). Effect of hyperthermia on the antiproliferative activities of murine α-, β-, and γ-interferon: Differential enhancement of murine γ-interferon. *Cancer Res.* **46**, 8–13.

Gerner, E.W., Holmes, D.K. and Stickney, D. (1980). Enhancement of hyperthermia-induced cytotoxicity by polyamines. *Cancer Res.* **40**, 432–438.

Gerner, E.W. and Schneider, M.J. (1975). Induced thermal resistance in HeLa cells. *Nature* **256**, 500–502.

Gerweck, L.E. (1977). Modification of cell lethality at elevated temperature. The pH effect. *Radiat. Res.* **70**, 224–235.

Gerweck, L.E., Jennings, M. and Richard, B. (1980). Influence of pH on the response of cells to single and split doses of hyperthermia. *Cancer Res.* **40**, 4019–4024.

Gerweck, L.E. (1985). Hyperthermia in cancer therapy: The biological basis and unresolved questions. *Cancer Res.* **45**, 3408–3414.

Goldenberg, D.M. and Langner, M. (1971). Direct and abscopal antitumor action of local hyperthermia. *Z. Naturforsch* **26b**, 359–361.

Goldin, E.M. and Leeper, D.B. (1981). The effect of low pH on the thermotolerance induction using fractionated 45 °C hyperthermia. *Radiat. Res.* **85**, 472–479.

Gullino, P.M., Grantham, F.H., Smith, S.H. and Haggerty, A.C. (1965). Modifications of the acid–base status of the internal milieu of tumors. *J. Natl. Cancer Inst.* **34**, 857–869.

Hahn, E.W., Alfieri, A.A. and Kim, J.H. (1979). The significance of local tumor hyperthermia/radiation on the production of disseminated disease. *Int. J. Radiat. Oncol. Biol. Phys.* **5**, 819–823.

Harris, J.W. (1976). Effects of tumor-like assay conditions, ionizing radiation, and hyperthermia on immune lysis of tumor cells by cytotoxic T-lymphocytes. *Cancer Res.* **36**, 2733–2739.

Hart, I.R. and Fidler, I.J. (1981). The implications of tumor heterogeneity for studies on the biology and therapy of cancer metastasis. *Biochim. Biophys. Acta* **651**, 37–50.

Henle, K.J. and Dethlefsen, L.A. (1978). Heat fractionation and thermotolerance: A review. *Cancer Res.* **38**, 1843–1851.

Henle, K.J., Karamuz, J.E. and Leeper, D.B. (1978). Induction of thermotolerance in Chinese hamster ovary cells by high (45 °) or low (40 °) hyperthermia. *Cancer Res.* **38**, 570–574.

Henle, K.J. (1980). Sensitization to hyperthermia below 43 °C induced in Chinese hamster ovary cells by step-down heating. *J. Natl. Cancer Inst.* **64**, 1479–1483.

Henle, K.J., Peck, J.W. and Higashikubo, R. (1983). Protection against heat-induced cell killing by polyols *in vitro*. *Cancer Res.* **43**, 1624–1627.

Heron, I. and Berg, K. (1978). The actions of interferon are potentiated at elevated temperature. *Nature* **273**, 508–510.

Hewitt, H.B. and Wilson, W. (1959). A survival curve of mammalian leukemia cells irradiated *in vivo*. (Implications for the treatment of mouse leukemia by whole body irradiation). *Br. J. Cancer* **13**, 68–75.

Hill, R.P. and Bush, R.S. (1969). A lung-colony assay to determine the radiosensitivity of the cells of a solid tumor. *Int. J. Radiat. Biol.* **15**, 435–444.

Hill, S.A. and Denekamp, J. (1978). The effect of vascular occlusion on the thermal sensitization of a mouse tumor. *Br. J. Radiol.* **51**, 997–1002.

Hill, S.A. and Denekamp, J. (1982). Does local tumor heating in mice influence metastatic spread? *Br. J. Radiol.* **55**, 444–451.

Hill, S.A. and Denekamp, J. (1982). Histology as a method for determining thermal gradients in heated tumors. *Br. J. Radiol.* **55**, 651–656.

Jahde, E. and Rajewsky, M.F. (1982). Tumor-selective modification of cellular microenvironment *in vivo*: Effect of glucose infusion on the pH in normal and malignant rat tissues. *Cancer Res.* **42**, 1505–1512.

Jahde, E., Rajewsky, M.F. and Baumgartt, H. (1982). pH distribution in transplanted neural tumors and normal tissues of BDIX rats as measured with pH microelectrodes. *Cancer Res.* **42**, 1498–1504.

Jain, R.K., Shah, S.A. and Finney, P.L. (1984). Continuous noninvasive monitoring of pH and temperature in rat Walker 256 carcinoma during normoglycemia and hyperglycemia. *J. Natl. Cancer Inst.* **73**, 429–436.

Joshi, D.S. and Jung, H. (1979). Thermotolerance and sensitization induced in CHO cells by fractionated hyperthermic treatments at 38–45 °C. *Eur. J. Cancer* **15**, 345–350.

Kahler, H. and Moore, B. (1962). pH of rat tumors and some comparisons with the Lissamine-Green circulation test. *J. Natl. Cancer Inst.* **28**, 561–568.

Kahler, H. and Robertson, W.B. (1943). Hydrogen-ion concentration of normal liver and hepatic tumors. *J. Natl. Cancer Inst.* **3**, 495–501.

Kase, K.R. and Hahn, G.M. (1976). Comparison of some response to hyperthermia by normal human diploid cells and neoplastic cells from the same origin. *Eur. J. Cancer* **12**, 481–491.

Kim, J.H., Kim, S.H., Hahn, E.W. and Song, C.W. (1978). 5-thio-D-glucose selectively potentiates hyperthermic killing of hypoxic tumor cells. *Science* **200**, 206–207.

Kim, J.H., Kim, S.H., Alfieri, A., Young, C.W. and Silvertrini, B. (1984). Lonidamine: A hyperthermic sensitizer of HeLa cells in culture and of the Meth-A tumor *in vivo*. *Oncology* **41**, Suppl. 1, 30–35.

Kim, S.H., Kim, J.H., Alfieri, A., He, S.O. and Young, C.W. (1985). Gossypol, a hyperthermic sensitizer of HeLa cells. *Cancer Res.* **45**, 6338–6340.

Law, M.P., Coultas, P.G. and Field, S.B. (1979). Induced thermal resistance in the mouse ear. *Br. J. Radiol.* **52**, 308–314.

Lechninger, A.L. (1981) *Biochemistry*. Worth Publishers, Inc. New York, p. 849.

Leith, J.T. (1982). Effect of methylglyoxal bis (quanylhydrazone) on skin reactions in the mouse after fractionated hyperthermic exposures. *Radiat. Res.* **90**, 586–594.

Li, G.C. and Mak, J.Y. (1985). Induction of heat shock protein synthesis in murine tumors during the development of thermotolerance. *Cancer Res.* **45**, 3816–3824.

Lord, P.F., Kapp, D.S. and Morrow, D. (1981). Increased skeletal metastases of spontaneous canine osteosarcoma after fractionated hyperthermia and local X-irradiation. *Cancer Res.* **41**, 4331–4334.

MacDonald, H.R. (1977). Effect of hyperthermia on the functional activity of cytotoxic T-Lymphocytes. *J. Natl. Cancer Inst.* **59**, 1263–1268.

Maher, J., Urano, M., Rick, L. and Suit, H.D. (1981). Thermal resistance in a spontaneous murine tumour. *Br. J. Radiol.* **54**, 1086–1090.

Marmor, J.B., Kozak, D. and Hahn, G.M. (1979). Effects of systemically administered bleomycin or adriamycin with local hyperthermia on primary tumor and lung metastases. *Cancer Treatment Reports*, **63**, No. 8, 1279–1290.

Milas, L., Hunter, N., Basia, I. and Withers, H.R. (1974). Complete regressions of an established murine fibrosarcoma induced by systemic application of *Corynebacterium granulosum*. *Cancer Res.* **34**, 2470–2475.

Miyakoshi, J. and Heki, S. (1983). Thermosensitization by methylglyoxal Bis (quanylhydrazone) in Chinese hamster cells. *Radiat. Res.* **96**, 523–531.

Mondovi, B., Santoro, A.S., Strom, R., Faiola, R. and Fanelli, A.R. (1972). Increased immunogenicity of Ehrlich ascites cells after heat treatment. *Cancer* **30**, 885–888.

Naeslund, J. and Swenson, K.E. (1953). Investigations on the pH of malignant tumours in mice and humans after the administration of glucose. *Acta Obstet. Gynecol. Scand.* **32**, 359–367.

Nielsen, O.S. and Overgaard, J. (1982). Importance of pre-heating temperature and time for the induction of thermotolerance in a solid tumour *in vivo*. *Br. J. Cancer* **46**, 894–903.

Oda, M., Koga, S. and Maeta, M. (1985). Effect of total-body hyperthermia on metastases from experimental mouse tumors. *Cancer Res.* **45**, 1532–1535.

Overgaard, J. and Bichel, P. (1977). The influence of hypoxia and acidity on the hyperthermic response of malignant cells *in vitro*. *Radiology* **123**, 511–514.

Overgaard, J. and Suit, H.D. (1979). Time-temperature relationship in hyperthermic treatment of malignant and normal tissue *in vivo*. *Cancer Res.* **39**, 3248–3253.

Reinhold, H.S. and van den Berg-Block, A. (1981). Enhancement of thermal damage to the microcirculation of 'Sandwich' Tumour by additional treatment. *Eur. J.Cancer Clin. Oncol.* **17**, 781–795.

Rhee, J.G., Song, C.W. and Levitt, S.H. (1982). Changes in thermosensitivity of mouse mammary carcinoma following hyperthermia *in vivo*. *Cancer Res.* **42**, 4485–4489.

Rhee, J.G., Kim, T.H., Levitt, S.H. and Song, C.W. (1984). Changes in acidity of mouse tumor by hyperthermia. *Int. J. Radiat. Oncol. Biol. Phys.* **10**, 393–399.

Rice, L.C., Urano, M. and Maher, J. (1982). The kinetics of thermotolerance in the mouse foot. *Radiat. Res.* **89**, 291–297.

Robinson, J.E., McCready, W.A., and Slawson, R.G. (1978). Thermal sensitivity of mouse mammary tumors. In *Cancer Therapy by Hyperthermia and Radiation*, Christian Streffer (Ed.), Proceedings of the 2nd International Symposium, Essen, June 2–4, 1977, Urban & Schwarzenberg, Baltimore, Munich, pp. 242–244.

Schrek, R. (1966). Sensitivity of normal and leukemic lymphocytes and leukemic myeloblasts to heat. *J. Natl. Cancer Inst.* **37**, No. 5, 649–654.

Sculier, J.P., Nicaise, C. and Klastersky, J. (1983). Lactic acidosis: A metabolic complication of extensive metastatic cancer. *Eur. J. Cancer Clin. Oncol.* **19**, 597–601.

Song, C.W., Clement, J.J. and Levitt, S.H. (1976). Preferential cytotoxicity of 5-thio-D-glucose against hypoxic tumor cells. *J. Natl. Cancer Inst.* **57**, 603–605.

Song, C.W., Guertin, D.P. and Levitt, S.H. (1979). Potentiation of cytotoxicity of 5-thio-D-glucose or hypoxic cells by hyperthermia. *Int. J. Radiat. Oncol. Biol. Phys.* **5**, 965–970.

Song, C.W., Kang, M.S., Rhee, J.G. and Levitt, S.H. (1980). The effect of hyperthermia on vascular function, pH, and cell survival. *Radiology* **137**, 795–803.

Song, C.W., Rhee, J.G. and Levitt, S.H. (1982). Effect of hyperthermia on hypoxic cell fraction in tumor. *Int. J. Radiat. Oncol. Biol. Phys.* **8**, 851–856.

Suit, H.D., Shalek, R.J. and Wett, R. (1965). Radiation response of C3H mouse mammary carcinoma evaluated in terms of cellular radiation sensitivity. In *Cellular Radiation Biology*, Williams and Wilkins Co., Baltimore, pp. 514–530.

Suit, H.D., Sedlacek, R.S. and Wiggins, S. (1977). Immunogenicity of tumor cells inactivated by heat. *Cancer Res.* **37**, 3836–3837.

Symonds, R.P., Wheldon, T.E., Clarke, B. and Bailey, G. (1981). A comparison of the response to hyperthermia of murine haemopoietic stem cells (CFU–S) and L1210 leukaemia cells: Enhanced killing of leukaemic cells in presence of normal marrow cells. *Br. J. Cancer* **44**, 682–691.

Thomlinson, R.H. and Gray, L.H. (1955). The histological structure of some human lung cancers and the possible implications for radiotherapy. *Br. J. Cancer* **9**, 539–549.

Tomasovic, S.P., Thames, H.D. and Nicolson, G.L. (1982). Heterogeneity in hyperthermic sensitivities of rat 13362 NF mammary adenocarcinoma cell clones of differing metastatic potentials. *Radiat. Res.* **91**, 555–563.

Urano, M. (1980). Long-term observation of mouse foot reaction after hyperthermia: hyperthermia may or may not be carcinogenic? *Br. J. Radiol.* **54**, 534–536.

Urano, M. (1986). The kinetics of thermotolerance in normal and tumor tissues. A review. *Cancer Res.* **46**, 474–482.

Urano, M. and Kahn, J. (1983). The change in hypoxic and chronically hypoxic cell fraction in murine tumors treated with hyperthermia. *Radiat. Res.* **96**, 549–559.

Urano, M. and Kahn, J. (1983). The effect of step-down heating on murine normal and tumor tissues. *Radiat. Res.* **94**, 350–358.

Urano, M. and Kahn, J. (1986). Differential kinetics of thermal resistance (thermotolerance) between murine normal and tumor tissues. *Int. J. Radiat. Oncol. Biol. Phys.* **12**, 89–93.

Urano, M. and Kahn, J. (1987) Some practical questions in the tumor regrowth assay. In *Rodent Tumor Models in Experimental Cancer Therapy*, Robert Kallman (ed.), Workshop in Reisenburg, West Germany, Pergamon Press, New York. pp. 122–127.

Urano, M., Gerweck, L.E., Epstein, R., Cunningham, M. and Suit, H.D. (1980). Response of spontaneous murine tumor to hyperthermia: Factors which modify the thermal response *in vivo*. *Radiat. Res.* **83**, 312–322.

Urano, M., Maher, J., Rice, L. and Kahn, J. (1982). Response of spontaneous murine tumor to hyperthermia: Temperature dependence in two different sized tumors. *Natl. Cancer Inst. Monograph* **61**, 299–301.

Urano, M., Montoya, V. and Booth, A. (1983). The effect of hyperglycemia on the thermal response of murine normal and tumor tissues. *Cancer Res.* **43**, 453–455.

Urano, M., Overgaard, M., Suit, H.D., Dunn, P. and Sedlacek, R. (1978). Enhancement by Corynebacterium parvum of the normal and tumor tissue response to hyperthermia. *Cancer Res.* **38**, 862–864.

Urano, M., Rice, L., Epstein, R., Suit, H.D. and Chu, A.M. (1983). Effect of whole-body hyperthermia on cell survival, metastasis frequency and host immunity in moderately and weakly immunogenic murine tumors. *Cancer Res.* **43**, 1039–1043.

Urano, M., Suit, H.D., Dunn, P., Lansdale, T. and Sedlacek, R. (1979). Enhancement of the thermal response of animal tumors by C. parvum. *Cancer Res.* **39**, 3454–3457.

Urano, M., Yamashita T, Suit, H.D. and Gerweck, L.E. (1984). Further studies on the enhancement of thermal response of normal and malignant tissues by *Corynebacterium parvum*. *Cancer Res.* **44**, 2341–2347.

Vaupel, P. (1982). Impact of localized microwave hyperthermia on the pH-distribution in malignant tumors. *Strahlentherapie* **158**, 168–173.

Voegtlin, C. (1936). Experimental studies on cancer I. The influence of the parenteral administration of certain sugars on the pH of malignant tumors. *Natl. Inst. Health Bull.* **164**, 1–14.

von Ardenne, M. (1980). Selective occlusion of cancer tissue capillaries as the central mechanism of the cancer multistep therapy. *Jpn. J. Clin. Oncol.* **10**, No. 1, 31–48.

von Ardenne, M. (1980). Hyperthermia and cancer therapy. *Cancer Chemother. Pharmacol.* **4**, 137–138.

von Ardenne, M., Chaplain, R.A. und Reitnauer, P.G. (1969). *In vivo* Versuche zue Krebs-Mehrschritt-Therapie mit der Attackenkombination Optimierte Tumorubersauerung + Hyperthermie + schwache Rontgenbestrahlung. *Das Deutsche Gesundheitswesen* **20**, 924–935.

Walker, A., McCallum, H.M., Wheldon, T.E., Nias, A.H.W. and Abdelaal, A.S. (1978). Promotion of metastasis of C3H mouse mammary carcinoma by local hyperthermia. *Br. J. Cancer* **38**, 561–563.

Wallen, C.A., Michaelson, S.M. and Wheeler, K.T. (1981). Temperature and cell survival variability across 9L subcutaneous tumors heated with microwaves. *Radiat. Res.* **85**, 281–291.

Wilk-Hooley, J.L., van den Berg, A.P., van der Zee, J., and Reinhold, H.S. (1984). The relevance of tumor pH to the treatment of malignant disease. Review Article. *Radiotherapy and Oncology* **2**, 343–366.

Woodruff, M.F.A., McBride, W.H. and Dunbar, N. (1974). Tumor growth, phagocytic activity and anti-body response in *Corynebacterium parvum*-treated mice. *Clin. Exp. Immunol.* **17**, 509–518.

Yatvin, M.B., Clifton, K.H. and Dennis, W.H. (1979). Hyperthermia and local anesthetics: Potentiation of survival of tumor-bearing mice. *Science* **205**, 195–196.

Yau, T.M. and Kim, S.C. (1980). Local anesthetics as hypoxic radiosensitizers, oxic radioprotectors and potentiators of hyperthermic killing in mammalian cells. *Br. J. Radiol.* **53**, 687–692.

Yerushalmi, A. (1976). Influence on metastatic spread of whole-body or local tumor hyperthermia. *Eur. J. Cancer* **12**, 455–463.

Yerushalmi, A., Tovey, M.G. and Gresser, I. (1982). Anti-tumor effect of combined interferon and hyperthermia in mice. *Proc. Soc. Exp. Biol. Med.* **169**, 413–415.

Hyperthermia and Oncology, Vol. 1, pp. 201–212 (1988)
Urano and Douple (Eds)
© 1988 VSP.

Chapter 7

Practical concepts of thermal dose

STEPHEN A. SAPARETO
Division of Hematology and Oncology, Department of Internal Medicine,
Wayne State University School of Medicine, Detroit, MI 48201, USA

A. INTRODUCTION

The ability to calculate some quantifiable description of thermal treatments is essential if hyperthermia is to become a routinely accepted and widely used modality for the treatment of cancer. However, *no acceptable unit of 'dose' which can be used to quantitate a therapeutic thermal treatment yet exists*. The solution to this problem will require major advances in both the measurement of temperature distribution as a function of time during hyperthermic treatment (i.e. three-dimensional temperature mapping) and in the conversion of these measurements into a prediction of treatment response. While the first of these requirements is beyond the scope of this discussion, the latter can be divided into two separate problems.

The first problem is to develop the ability to quantitate the temperature as a function of time for any *single point* within the volume of interest into a number representative of the 'dose' of heat at that point. The second problem is to develop a means of converting this three-dimensional 'dose' map into a *clinically useful* description of the hyperthermia treatment which can be correlated with response. Clearly, it is essential to solve the first of these problems before addressing the second.

While it may be theoretically satisfying to define a thermal dose in terms of a physically quantifiable unit, progress in this direction has been very arduous. This is primarily due to a lack in understanding of the fundamental mechanisms of thermally-induced killing. Once the process leading to thermal death has been identified, it may be possible to elucidate the thermodynamic mechanisms involved. However, at present, the difficulty in developing such an approach lies in the fact that any unit must provide the ability to show a relationship with biological response. Clearly, both *temperature and time* must be related to this unit.

In order to satisfy the criteria of demonstrating a dose response, a variety of ideas have been proposed. Consequently, all of the approaches are, to some extent, empirical in nature. Some of the methods are based on adaptation of existing units and some propose new unique units.

B.   PROPOSED METHODS OF CALCULATING THERMAL DOSE

1.   Early efforts

A number of early attempts were made to describe hyperthermia treatments based on existing physical units. Among these are: the total heat energy transferred to the tissue (Pettigrew *et al.* 1974); the duration of exposure above some selected temperature (Stehlin Jr. *et al.* 1975); and highest temperature (LaVeen *et al.* 1976) or lowest temperature (Hahn and Boone 1976) achieved. However, all of these approaches are lacking in one way or another.

The influence of radiation therapy has suggested the use of energy deposition as a dose unit for heat. However, hyperthermia differs greatly from radiation in that a simple measurement of energy deposition, such as the Gray, cannot be used to demonstrate any dose response for thermal damage. Clearly, it is not solely the rise in temperature that causes biological damage, but rather the time spent at that elevated temperature that provides a dose response relationship. Thus, any quantitation of a thermal treatment which is to provide a dose response relationship must be a function of both temperature and time. Furthermore, the existing literature has demonstrated that biological effects are not a linear function of temperature and time, but rather they are an exponential function of these two factors (see reviews by Henle and Dethlefsen 1980; Field and Morris 1983).

The use of a target temperature for a fixed period of time is a common convention in clinical hyperthermia. Usually time measurement begins when a monitoring probe achieves an arbitrarily sufficient preselected temperature. However, this method has difficulties. First, it neglects the contribution of the warm-up period to the treatment. Second, and perhaps even more important, it often is not possible to achieve or maintain a pre-chosen temperature over a given treatment volume due to such limitations as patient discomfort, insufficient power, tumor location, tumor vasculature or a combination of these factors. Thus, this method does not allow for comparison of treatments at different temperatures.

To date, the majority of efforts in developing a useful quantitation of thermal dose has relied on an empirical evaluation of the biological dose response. The description of dose in terms of biological response is quite common, for example, in the dosage of antibiotics in standard units of biological activity. Since most methods based upon the use of biological response to define a unit of thermal dose, in essence, involve the use of iso-effects to determine a mathematical description, such methods may be termed *thermal iso-effect dose calculations*. The majority of the proposed methods are similar in that they essentially describe an exponential relationship between time and temperature for a given effect.

2. Degree-minutes

The simplest description of temperature and time is an integration of the temperature above a given initial temperature times the time spent at that temperature. This is nothing more than the area under the time versus temperature curve, analogous to the concentration times time method used in chemotherapy, and assumes a linear relationship between time and temperature.

Pettigrew *et al.* (1974a) used 'degree-hours' above a given temperature as a means of describing thermal treatment for whole-body hyperthermia and Dewhirst *et al.* (1984) have evaluated data for local hyperthermia of spontaneous tumors in dogs and cats by this approach. However, while this approach is straightforward, it conflicts with the clear observation that the time–temperature relationship is exponential.

3. Fractional unit of cell kill

Atkinson (1977) proposed that a unit, F, the fractional cell kill could be used as a unit of thermal dose by the definition:

$$F = 100 - 100e^{-\int a(t)dt} \tag{1}$$

Where $a = \alpha\, e^{E/kT}$, α is a temperature independent rate constant, E is the activation energy, and k is the Boltzmann constant.

This approach, in essence, is derived from an assumption of simple first order rate kinetics for thermal killing and attempts to describe the shape of the *in vitro* cell survival curve. However, while this approach was one of the first to apply basic thermodynamic theory to cellular inactivation, the proposed units tend to approach a maximum value (100) at times less than one hour at 43 °C, thus limiting its usefulness in practical therapy.

4. Arizona Standard Hyperthermia (ASH) Unit

Gerner (1983) proposed a dose unit related to the approach of Atkinson which he termed the ASH unit:

$$\text{ASH Dose} = \int \frac{T(t)\, e^{-\frac{[\Delta H^+ + RT(t)]}{[RT(t)]}}}{h(\Delta H^+)}\, dt \tag{2}$$

Where ΔH^+ is the inactivation enthalpy, R is the universal gas constant, and $h(\Delta H^+) = 10^{-100}$ is used as a scaling factor. This method also is derived from a simple application of first order rate kinetics, obtaining a enthalpy value from an empirical fit of *in vitro* survival data. However, although, more practical for therapeutic use, it presents several difficulties including the lack of accounting for the change in the enthalpy value which is known to occur at about 43 °C and the troublesome use of a scaling factor. Gerner has extended his efforts toward a more fundamental approach related to thermodynamics and the apparent Gibbs free energy of cellular inactivation

(Gerner 1985). However, as of yet, he has not proposed a practical unit from this work.

5. Basal metabolic dose

Robbins *et al.* (1985) have proposed a unique thermal dose unit for whole-body hyperthermia based on the human basal metabolic rate (BMR). They define a dose unit as an arbitrarily chosen 20 watt increase in basal metabolic rate. Since the BMR changes as a function of temperature by 7 per cent for each 0.5 °C (Law and Pettigrew 1980), this dose unit can be defined for a person whose BMR at 37 °C is approximately 85 watts as:

$$\text{BMR dose} = \frac{t \times 85 \times 1.07^{(T-37)/0.5}}{20 \times 60} \qquad (3)$$

The rationale for this approach has not been described. However, note that the unit exponentially relates time and temperature in a manner similar to the majority of other units described here.

6. Equivalent minutes at 43 °C

The relationship between temperature and exposure time during hyperthermic treatments has been reported for a variety of biological systems. The evidence clearly shows that, for both *in vitro* and *in vivo* systems, an exponential relationship exists between temperature and exposure time. *This relationship appears general in that it holds for almost all of the biological systems that have been studied* (Suit and Schwayder 1974; Dewey *et al.* 1977; Field 1978; Overgaard and Suit 1979; Henle and Dethlefsen 1980; and Field and Morris 1983). Simply stated: a one degree increase in temperature requires a two-fold decrease in time for the same effect above 43 °C and a 3–to–6 fold decrease in time for an iso-effect below 43 °C. Based on this description, a simple thermal iso-effect calculating method (referred to as EQ43) has been described (Dewey *et al.* 1976; Sapareto and Dewey, 1984). For ease in calculation, the relationship may be described as a numerical integration:

$$\text{EQ43} = \sum_{t=0}^{t=\text{final}} R^{(43-\bar{T})} \Delta t \qquad (4)$$

where $\bar{T}$ is the average temperature during time Δt, $R = 0.5$ above 43 °C and $R = 0.25$ below 43 °C.

The utility of EQ43 calculations is demonstrated in Fig. 1 where the *in vitro* survival of cells exposed to various temperatures are plotted as as function of equivalent-minutes at 43 °C (Sapareto *et al.* 1978). The data fit

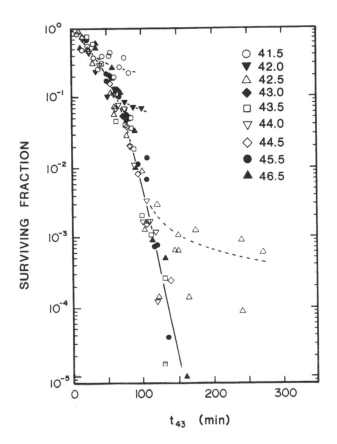

Figure 1: Survival of Chinese hamster ovary cells as a function of EQ43 dose for various temperatures using 43 °C as the transition temperature and *R* values above and below the transition of 0.5 and 0.25, respectively. Reproduced with permission from Sapareto and Dewey (1984).

a single curve ($r^2 = 0.87$) with the notable exception of extended heating at temperatures of 42.5 °C and below. At these temperatures, survival deviates from the single curve presumably due to the development of increasing thermotolerance during the heat treatment. Since the rate of the development of thermotolerance is probably temperature dependent (Leeper 1983), most likely the assumption of a constant value for *R* below 43.0 °C is not valid. The fit of data below the transition temperature is highly sensitive to the temperature chosen for the transition and the *R* value below that transition. Figure 2 demonstrates how the fit for the same data as Fig. 1, using a transition temperature of 42.5 °C with no change in the *R* values, shows even greater deviation from the data above 43 °C. Clearly, an important consideration for appropriate use of this dose method is the selection of the best values for these parameters and further efforts are necessary to devise a better means to account for the phenomenon of thermotolerance.

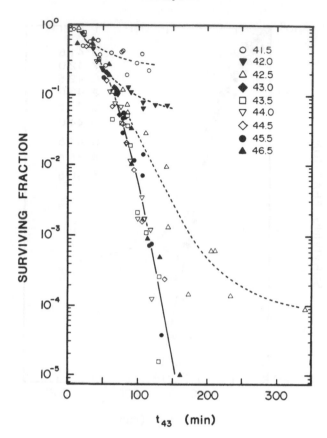

Figure 2: Survival of Chinese hamster ovary cells as a function of EQ43 dose for various temperatures using 42.5 °C as the transition temperature and R values above and below the transition of 0.5 and 0.25, respectively.

The relationship described by the EQ43 calculation provides a useful method to calculate the accumulated dose at a reference temperature under a variety of heating profiles, including those temperature histories that cannot be easily described mathematically. However, it has evolved primarily from studies using single uniform doses of heat. Clearly, this situation is not likely in the clinic. The primary goal of future efforts must be to improve these thermal dose calculations to provide accurate assessment of treatment on complex (i.e. more clinically relevant) temperature histories *in vivo* for application to human therapy.

7.   Variations in equivalent-minutes calculations

Field and Morris (1983), after careful evaluation of the available *in vivo* data, have proposed that the transition temperature should be at 42.5 °C and that the R value below this transition temperature should be 0.167 (a factor of 6 per degree instead of 4). Figure 3 is a replot of Fig. 1 using these new values. This curve shows only the data points below 43 °C, since only this data is affected. As can be seen, the data fit the optimum curve above 43 °C (from Fig. 1) much more poorly at low doses, cross the curve at higher doses, and fit poorly again at still higher doses. One possible explanation for this discrepancy is that the endpoints used in the *in vivo* data studied by Field and Morris are all relatively severe effects (e.g., tumor cure, foot loss, ear necrosis, etc.). Therefore, it is possible that their analysis has optimized the transition temperature and R value for equivalence to the data above 43 °C (the line) at high doses at the expense of a poor fit at low doses. Since it is likely that clinical thermal doses frequently will be below an EQ43 dose of 100 minutes (Dewhirst *et al.* 1984), the accuracy of EQ43 at low doses clearly is important. Indeed, Field and Morris (1983) also demonstrate how minor changes in the transition temperature and R value

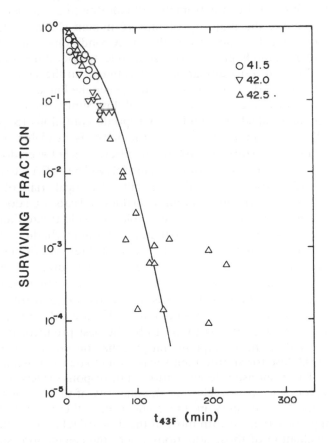

Figure 3: Survival of Chinese hamster ovary cells as a function of EQ43 dose for various temperatures using 42.5 °C as the transition temperature and R values above and below the transition of 0.5 and 0.165, respectively. Data at temperatures above 42.5 °C have been eliminated for clarity since they are not affected by this change in values. The line is from Fig. 1.

can significantly alter the calculated thermal dose, lending further emphasis to the need for more study of this difficulty.

Sapozink (1986) also has proposed a modification of the equivalent-minutes calculation as described above. For computational simplicity, he constrains $R^{(T-43)}$ to 0 below 40 °C and 64 above 49 °C, however, there is no theoretical rationale for these limits. While these boundary conditions may, in most cases, cause insignificant changes in the calculated thermal dose, there are some instances where differences in calculation could occur such as a brief, transient excursion above 49 °C in a critical normal tissue, or in highly heat sensitive tissues (such as that sensitized by low pH) where some effect might be seen even below 40 °C.

C. CLINICAL EVALUATION

The evaluation of thermal dose calculations as a prognostic indicator of treatment response has been extremely difficult due to a number of factors. First, and foremost, the majority of mature human clinical data contains limited thermometry data. Second, under many conditions, since all practical dose calculating methods give exponential time–temperature relationships, they would all tend to rank any group of treatments in the same order, thus demonstrating similar dose responses. In fact, even the linear degree-minutes calculations can provide estimates of thermal dose similar to the exponential equivalent-minutes calculation under appropriate conditions (Sapareto and Dewey 1984). Thus, until adequate thermometry is available and clinical trials can be designed to directly investigate differences between dose models, any significant conclusions about different models will be difficult to achieve.

Dewhirst et al. (1984) evaluated the results of clinical trials of combined heat and radiation on spontaneous tumors in dogs and cats for a dose response with hyperthermia treatment. They were able to divide treated animals into quartiles based on increasing dose and then to determine whether a correlation with tumor response could be seen. Since at least three and as many as eight thermal probes were used for each treatment and measurements were taken at least every 5 minutes, it was possible to compare minimum, maximum, average and range of thermal dose measurements for each treatment with response. In evaluating degree-minute and EQ43 doses, the minimum EQ43 dose on the first treatment was found to be the best predictor of long-term response ($P < 0.05$). On subsequent analysis, the non-probe specific average minimum EQ43 dose for all treatments was found to be a consistent predictor of both complete response rate and duration of response (Dewhirst and Sim 1984). However, since it is likely that the majority of thermal dose methods might rank these treatments in the same order, these correlations with response might be expected with most of the dose calculation methods. Only by careful evaluation of those situations where the calculations significantly differ could one attempt to compare different calculation methods.

Dewhirst and Sim (1984) further noted that the proportion of the tumor achieving a significant thermal dose also related to response rate and duration

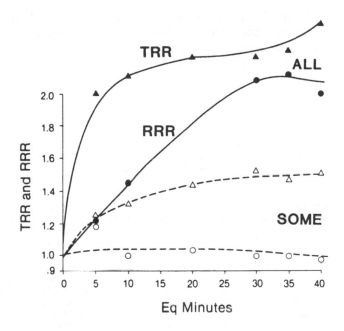

Figure 4: The influence of the proportion of the tumor heated versus a measure of the thermal enhancement of complete response (TRR) and duration of response (RRR) reprinted with permission from Dewhirst and Sim (1984). The closed symbols indicate the TRR (▲) and RRR (●) when all temperature probes were above the EQ43 dose indicated. The open symbols represent the results when only 'some' of the probes achieved the indicated dose. Both TRR and RRR are severely reduced if any of the probes do not get a significant thermal dose. (Dewhirst *et al.* 1983 for complete definitions of TRR and RRR.)

of response. In a comparison of tumors where either all of the treatment monitoring probes were above a given EQ43 dose or only some of the treatment probes were above the same EQ43 dose, both complete response rate and duration of response were found to be significantly greater when all of the treatment monitoring probes were above a given dose (Fig. 4).

The utility of thermal dose to predict normal tissue response was evaluated in a comparison of normal tissue complications on the same data (Dewhirst and Sim 1986). In this case, they noted that the maximum intratumor temperature correlated strongly with both the incidence and severity of normal tissue complications. They suggested that these maxima may, in fact, be related to excessive temperatures in adjacent normal tissues which were not extensively monitored.

In order to evaluate different methods of thermal dose for human clinical trials, a computer program has been developed to calculate several of the thermal doses described here. In order to better compare each dose unit, some units (degree-minute dose and basal metabolic dose) have been converted into times at 43 °C which give the same dose. The program is written in PASCAL for the IBM-PC and is designed to read data files produced by the Clini-therm Mark IV hyperthermia treatment system (Clini-therm Corp.,

Dallas, TX). A sample of the output from this program is shown in Table 1. The program has been freely distributed to interested hyperthermia treatment clinics which are using Clini-therm systems to provide a means to evaluate and compare different calculations.

Table 1.
Sample output from thermal dose calculating program

Report Date: 08/15/84 (12:27)   TCLIN 2.0
Report Type: Original   For Evaluation of Thermal Doses   File Name: B:TEST.001
Patient Name: Test, Ima   (C) S. Sapareto   Physician: Rufus T. Firefly
                                       Written by A. Horwitz

Probe	Sensor location	t_{43} (min)	(t_{43C}) (min)	T_{43F} (min)	t_{bmr} (min)	D_{bmr}	t_{dm43} (min)	T_{max} (°C)
1		0.0	(0.0)	0.0	3.4	(655.6)	0.0	22.2
2		0.0	(0.0)	0.0	3.7	(706.4)	0.0	22.3
3		0.0	(0.0)	0.0	4.6	(876.6)	0.0	24.6
4		0.0	(0.0)	0.0	3.7	(707.0)	0.0	22.4
5*	tumor, right ant	329.1	(330.0)	330.3	77.9	(14915.4)	75.9	46.8
6*	tumor, left post	2.7	(2.7)	2.2	43.8	(8389.1)	34.6	41.9
7*	tumor, left ant	11.9	(12.0)	14.2	50.8	(9727.6)	45.9	42.8
8		0.0	(0.0)	0.0	3.7	(711.0)	0.0	22.4
9*	tumor, right post	100.5	(106.1)	101.1	59.7	(11427.4)	55.3	47.6
10*	tumor, " "	539.2	(549.2)	541.1	75.2	(14401.1)	71.9	49.7
11*	tumor, " "	304.6	(308.2)	305.5	77.5	(14832.6)	76.4	48.2
12*	tumor, " "	142.4	(144.6)	144.3	66.9	(12814.1)	65.6	47.5

t_{43} =   Equivalent minutes at 43 °C (Sapareto and Dewey 1984) also Eq43 (Dewhirst *et al.* 1984)
t_{43F} =   Equivalent minutes at 43 °C (Field and Morris 1983)
t_{43C} =   Equivalent minutes from Clinitherm Mark IV System (Sapareto and Dewey 1984)
t_{dm43} =   Minutes at 43 °C for identical degree-minutes above 37 °C (Sapareto and Dewey 1984)
t_{bmr} =   Minutes at 43 °C for identical D_{bmr}
D_{bmr} =   Basal metabolic dose (Robins *et al.* 1985)

D. PROBLEMS AND FUTURE DIRECTIONS

Tumors are likely to show large differences in their sensitivity to heat because of variations in such factors as blood flow and pH. While this presents technical challenges, it does not negate the use of hyperthermia, but rather, emphasizes the central importance of estimating the thermal dose. An EQ43 dose calculation would be useful for two reasons. First, as in radiation treatments, the tissue which is dose limiting for curative hyperthermia treatments is the normal tissue within the treatment volume. Normal tissue is less likely to show variability due to the above mentioned factors. Thus it is essential that future testing of thermal dose calculations be performed in normal tissue models as well as tumor models. Second, the purpose of any thermal dose unit is not to account for variation in sensitivity of any specific tissue, whether normal or malignant; rather, a thermal dose unit should be used to quantitate these variations so that they may be studied and compared.

An analogous situation in radiation therapy is the difference between radioresistant and radiosensitive tumors, where different tumors exhibit different radiation sensitivities presumably due to variability in tumor hypoxia and repair capabilities. In this case, based on the changes noted in the dose response curves, oxygen has been found to be a dose modifying factor leading to the concept of the oxygen enhancement ratio (OER). In this manner, such known complications as step-down heating and thermotolerance can be studied.

A thermal dose model which accurately predicts response for normal tissue would provide a method for determining whether tumors are more or less sensitive to heat under various protocols. This would be of great benefit in predicting therapeutic gain. For example, if dose-equivalent normal tissue treatments at different temperatures could be given to tumors, one could assess the question of whether an optimum hyperthermic temperature for therapeutic gain exists. A temperature which gives the maximum tumor response under these conditions would be the temperature for maximum therapeutic gain.

Clearly, the need for further research into methods of predicting thermal dose is evident. While the application and evaluation of human clinical hyperthermic therapy must await significant improvements in temperature measurement and prediction, efforts in better controlled systems both *in vitro* and *in vivo* can and should proceed.

REFERENCES

Atkinson, E.R. (1977). Hyperthermia dose definition. *J. Bioengng.* **1**, 487–492.
Dewey, W.C., Hopwood, L.E., Sapareto, S.A. and Gerweck, L.E. (1977). Cellular responses to combinations of hyperthermia and radiation. *Radiology* **123**, 463–479.
Dewhirst, M.W. and Sim, D.A. (1984). The utility of thermal dose as a predictor of tumor and normal tissue responses to combined radiation and hyperthermia. *Cancer Res.* **44**, 4772–4780 (Suppl.).
Dewhirst, M.W. and Sim, D.A. (1986). Estimation of therapeutic gain in clinical trials involving hyperthermia and radiotherapy. *Int. J. Hyperthermia* **2**, 165–178.
Dewhirst, M.W., Sim, D.A., Sapareto, S.A. and Connor, W.G. (1984). The importance of minimum tumor temperature in determining early and long-term responses of spontaneous pet animal tumors to heat and radiation. *Cancer Res.* **44**, 43–50.
Dewhirst, M.W., Sim, D.A., Wilson, S., DeYoung, D. and Parsells, J. (1983). Correlation between initial and long-term responses of spontaneous pet animal tumors to heat and radiation or radiation alone. *Cancer Res.* **43**, 5735–5741.
Field, S.B. (1978). The response of normal tissues to hyperthermia alone or in combination with X-rays. In *Cancer Therapy by Hyperthermia and Radiation*. Proceedings of the 2nd International Symposium, pp. 37–48, C. Streffer (Ed.), Urban and Schwarzenberg, Munich.
Field, S.B. and Morris, C.C. (1983). The relationship between heating time and temperature: its relevance to clinical hyperthermia. *Radiotherapy and Oncology* **1**, 179–186.
Gerner, E.W. (1983). Thermal dose: Definition and evaluation in terms of hyperthermia-induced cytotoxicity (abstract). *7th Int. Congress of Radiat. Res.*, Amsterdam, Netherlands.
Gerner, E.W. (1985). Thermal dose for hyperthermia-induced cytotoxicity. In *Proceedings of the Seventh Annual Conference of the IEEE Engineering in Medicine and Biology Society*. pp. 51–54. Lin, J.C. and Feinberg, B.N. (Ed.), IEEE Service Center, Piscataway, N.Y.
Hahn, G.M. and Boone, M.L. (1976). Heat in tumor therapy. *JAMA* **236**, 2286–2287.
Henle, K.J. and Dethlefsen, L.A. (1980). Time-temperature relationships for heat-induced killing of mammalian cells. *Ann. N.Y. Acad. Sci.* **335**, 234–253.

Law, H.T. and Pettigrew, R.T. (1980). Heat transfer in whole-body hyperthermia. *Ann. N.Y. Acad. Sci.* **335**, 298–310.

Leeper, D.B. (1983). Temperature dependence of thermotolerance induction in Chinese hamster ovarian carcinoma cells *in vitro* (abstract). *3rd Annual Meeting of the North American Hyperthermia Group*, San Antonio, TX.

LeVeen, H.H., Wapnick, V., Picone, V., Folk, G. and Ahmed, N. (1976). Tumor eradication by radiofrequency therapy. Response in 21 patients. *JAMA* **235**, 2178–2200.

Mendecki, J., Friedenthal, E. and Botstein, C. (1976) Effects of microwave-induced local hyperthermia on mammary adenocarcinoma in C3H mice. *Cancer Res.* **36**, 2113–2114.

Overgaard, J. and Suit, H.D. (1979). Time-temperature relationship in hyperthermic treatment of malignant and normal tissue *in vivo. Cancer Res.* **39**, 3248–3253.

Pettigrew, R.T., Galt, J.M., Ludgate, C.M., Horne, D.B. and Smith, A.N. (1974a). Circulatory and biochemical effects of whole-body hyperthermia. *Br. J. Surg.* **61**, 727–730.

Pettigrew, R.T., Galt, J.M., Ludgate, C.M. and Smith, A.N. (1974). Clinical effects of whole-body hyperthermia in advanced malignancy. *Br. Med. J.* **4**, 679–682.

Robbins, H.I., Dennis, W.H., Neville, A.J., Schecterle, L.M., Martin, P.A., Grossman, J., Davis, T.E., Neville, S.R., Gillis, W.K. and Rusy, B.F. (1985). A nontoxic system for 41.8 °C whole-body hyperthermia: Results of a phase I study using a radiant heat device. *Cancer Res.* **45**, 3937–3944.

Sapareto, S.A. and Dewey, W.C. (1984). Thermal dose determination in cancer therapy. *Int. J. Radiat. Oncol. Biol. Phys.* **10**, 787–800.

Sapareto, S.A., Hopwood, L.E., Dewey, W.C., Raju, M.R. and Gray, J.W. (1978). Hyperthermic effects on survival and progression of CHO cells. *Cancer Res.* **38**, 393–400.

Sapozink, M.D. (1986). The application of thermal dose in clinical trials. *Int. J. Hyperthermia* **2**, 157–164.

Stehlin Jr., J.S., Giovanella, B.C., de Ipolyi, P.D., Muenz, L.R. and Anderson, R.F. (1975). Hyperthermic perfusion with chemotherapy for cancers of the extremities. *Surg. Gynec. Obstet.* **140**, 338–349.

Suit, H.L. and Schwayder, M. (1974). Hyperthermia, potential as an anti-tumour agent. *Cancer* **34**, 122–129.

Subject index